Student Solutions Manual

for Blei and Odian's

General, Organic, and Biochemistry: Connecting Chemistry to Your Life.

Mark D. Dadmun

The University of Tennessee

W. H. Freeman and Company
New York

ISBN 0-7167-3579-2 (EAN: 9780716735793)

© 2000 by W. H. Freeman and Company. All rights reserved.

No part of this book may be reproduced by any mechanical, photographic, or electronic process, or in the form of a phonographic recording, nor may it be stored in a retrieval system, transmitted, or otherwise copied for public or private use, without written permission from the publisher.

Printed in the United States of America

Fourth printing

Table of Contents

Chapter 1	The Language of Chemistry	1
Chapter 2	Atomic Structure	7
Chapter 3	Molecules and Chemical Bonds	10
Chapter 4	Chemical Calculations	18
Chapter 5	The Physical Properties of Gases	26
Chapter 6	Interactions Between Molecules	32
Chapter 7	Solutions	35
Chapter 8	Chemical Reactions	42
Chapter 9	Acids, Bases, and Buffers	45
Chapter 10	Chemical and Biological Effects of Radiation	52
Chapter 11	Saturated Hydrocarbons	55
Chapter 12	Unsaturated Hydrocarbons	67
Chapter 13	Alcohols, Phenols, Ethers, and Their Sulfur Analogues	82
Chapter 14	Aldehydes and Ketones	92
Chapter 15	Carboxylic Acids, Esters, and Other Acid Derivatives	104
Chapter 16	Amines and Amides	118
Chapter 17	Stereoisomerism	130
Chapter 18	Carbohydrates	140
Chapter 19	Lipids	149
Chapter 20	Proteins	157
Chapter 21	Nucleic Acids	166
Chapter 22	Metabolism and Enzymes: An Overview	172
Chapter 23	Carbohydrate Metabolism	174

Chapter 24	Fatty Acid Metabolism	177
Chapter 25	Amino Acid Metabolism	180
Chapter 26	Hormones and the Control of Metabolic Interrelations	182
Appendix	Solutions to *Organic and Biochemistry* Chapters 1 and 2	A-1

Preface

Chemistry plays an important role in many aspects of everyday life, from cooking to personal care products to health care. Understanding how chemical transformations occur provides you with knowledge that can be called upon and utilized in the future to make smart consumer decisions, to fix problems, and to perform important job-related tasks. This last aspect is particularly true if you are planning a profession in the health sciences—a field in which an understanding of how medications work and how the body functions, *from a chemical standpoint,* is essential for a successful and prosperous career.

General, Organic, and Biochemistry by Blei and Odian is designed to provide an introduction to the concepts of chemistry for students who are planning a career in the health sciences. The book is separated into three sections corresponding to the title: Chapters 1–10 introduce the important concepts of general chemistry, Chapters 11–17 provide the fundamentals of organic chemistry, and Chapters 18–26 explain the chemistry of the body—biochemistry.

This manual provides the complete solutions to the odd-numbered exercises at the end of each chapter of *General, Organic, and Biochemistry*. The explanations were written to show you the thought process required to solve the problem. Additionally, the explanations exemplify the mathematical level and problem-solving skills that are required for mastery of the material in that chapter.

Although the answers to these questions can be found in the back of the text, the purpose of this manual is to explain *how* the answers are derived. It is important to use this manual as a check of your work. Though it may be tempting to peruse the manual before working the problem, you will only learn the material (which is the goal, isn't it?) if you work the problem first, and then refer to the manual.

This manual also can be used for *Organic and Biochemistry,* by Blei and Odian. This sister text synopsizes the concepts of general chemistry into two chapters and then proceeds to cover the important concepts of organic chemistry and biochemistry. It is considered a corresponding text to *General, Organic, and Biochemistry* because Chapters 1 and 2 summarize material from Chapters 1–10 of *General, Orgnaic, and Biochemistry*, and Chapters 3–18 correspond directly to Chapters 11–26 of *General, Organic, and Biochemistry.* Thus, Chapters 11–26 of this manual also provide the worked solutions to the odd-numbered end-of-chapter exercises for Chapters 3–18 of *Organic and Biochemistry.* The solutions to all end-of-chapter exercises in Chapters 1 and 2 of *Organic and Biochemistry* are provided in an Appendix at the end of this manual.

Finally, I am indebted to a number of people who have helped in the preparation of this manual. Most importantly, however, I would like to thank my family, Jayne, Ryan, and Catherine, for their love, patience, and support during the writing of this manual.

Chapter 1

The Language of Chemistry

1.1 A physical property is one that a substance can undergo without changing its identity. A chemical property is one that describes how a substance can react with another substance to form a new substance.

(a) Oxygen loses its identity; therefore, this is a chemical property.

(b) When oxygen boils, it transforms from a liquid into a gas but does not change identity; therefore, this is a physical property.

(c) When oxygen dissolves in water, it mixes with water on a molecular level but does not change identity; therefore, this is a physical property.

(d) When oxygen reacts with hydrogen to form water, oxygen loses its identity; therefore, this is a chemical property.

1.3 (a) Air is a mixture of oxygen and nitrogen.

(b) Mercury is a pure substance found in the periodic table.

(c) Aluminum foil is made only of aluminum; therefore, it is a pure substance.

(d) Table salt is NaCl, a pure compound that cannot be separated.

1.5 Length is measured in meter; mass in kilograms; temperature in Kelvins; time in seconds; amount of substance in moles.

1.7 (a) $630 \text{ m} \times \left(\dfrac{1 \text{ km}}{1000 \text{ m}}\right) = 0.630 \text{ km}$

(b) $1440 \text{ ms} \times \left(\dfrac{1 \text{ s}}{1000 \text{ ms}}\right) = 1.440 \text{ s}$

(c) $0.000065 \text{ kg} \times \left(\dfrac{1000 \text{ g}}{1 \text{ kg}}\right) \times \left(\dfrac{1000 \text{ mg}}{1 \text{ g}}\right) = 65 \text{ mg}$

(d) $1300 \text{ μg} \times \left(\dfrac{1 \text{ mg}}{1000 \text{ μg}}\right) = 1.3 \text{ mg}$

1.9 (a) $0.0945 = 9.45 \times 10^{-2}$; therefore three significant figures.

(b) $83.22 = 8.322 \times 10^1$; therefore four significant figures.

(c) $106 = 1.06 \times 10^2$; therefore three significant figures.

(d) $0.000130 = 1.30 \times 10^{-4}$; therefore three significant figures.

1.11 (a) 25.9000 g; because last three zeros are shown, all digits are significant; therefore six significant figures.

(b) 102 cm = 1.02×10^2 cm; therefore three significant figures.

(c) 0.002 m = 2×10^{-3} m; therefore one significant figure.

(d) 2001 kg = 2.001×10^3 kg; therefore four significant figures.

(e) 0.0605 s = 6.05×10^{-2} s; therefore three significant figures.

1.13 (a) 0.00839; move decimal point to the right three places to get 8.39; therefore ≡ 8.39×10^{-3}.

(b) 83,264; move decimal point to the left four places to get 8.3264; therefore ≡ 8.3264×10^4.

(c) 372; move decimal point to the left two places to get 3.72; therefore ≡ 3.72×10^2.

(d) 0.0000208; move decimal point to the right fve places to get 2.08; therefore ≡ 2.08×10^{-5}.

1.15 (a) 936,800; move decimal point to the left five places to get 9.368; therefore ≡ 9.368×10^5.

(b) 1638; move decimal point to the left three places to get 1.638; therefore ≡ 1.638×10^3.

(c) 0.0000568; move decimal point to the right five places to get 5.68; therefore ≡ 5.68×10^{-5}.

(d) 0.00917; move decimal point to the right three places to get 9.17; therefore ≡ 9.17×10^{-3}.

1.17 (a) In multiplication, the number of significant figures in the answer is equal to the number of significant figures in the least exact starting number. In this example, 3.21 has three significant figures; therefore the answer must have three significant figures ≡ 48.4 cm^2.

(b) 1.1 has two significant figures; therefore the answer must have two significant figures ≡ 8.4.

1.19 In the addition of two numbers, the result should have only as many digits to the right of the decimal point as the starting number with the least number of digits to the right of the decimal point has.

(a) 1.23 has only two digits after the decimal point; therefore 1.23 + 12.786 = 14.02.

(b) 3.961 has only three digits past the decimal point; therefore 3.961 + 24.6543 = 28.615.

1.21 Each answer will have three significant figures because 0.800 miles has three significant figures, which is a minimum.

(a) $0.800 \text{ miles} \times \left(\dfrac{1.6093 \text{ km}}{1 \text{ mile}}\right) = 1.29 \text{ km}$

(b) $0.800 \text{ miles} \times \left(\dfrac{1.6093 \text{ km}}{1 \text{ mile}}\right) \times \left(\dfrac{1000 \text{ m}}{1 \text{ km}}\right) = 1.29 \times 10^3 \text{ m}$

(c) $0.800 \text{ miles} \times \left(\dfrac{1.6093 \text{ km}}{1 \text{ mile}}\right) \times \left(\dfrac{1000 \text{ m}}{1 \text{ km}}\right) \times \left(\dfrac{100 \text{ cm}}{1 \text{ m}}\right) = 1.29 \times 10^5 \text{ cm}$

(d) $0.800 \text{ miles} \times \left(\dfrac{1.6093 \text{ km}}{1 \text{ mile}}\right) \times \left(\dfrac{1000 \text{ m}}{1 \text{ km}}\right) \times \left(\dfrac{1000 \text{ mm}}{1 \text{ m}}\right) = 1.29 \times 10^6 \text{ cm}$

1.23 $3.75 \text{ gal} \times \left(\dfrac{3.7853 \text{ L}}{1 \text{ gal}}\right) \times \left(\dfrac{1000 \text{ mL}}{1 \text{ L}}\right) = 1.42 \times 10^4 \text{ mL}$

1.25 $10 \text{ km} \times \left(\dfrac{1 \text{ mile}}{1.6093 \text{ km}}\right) = 6.214 \text{ miles}$

1.27 From Table 1.5: 1 oz = 28.35 g

1.29 (a) $0.60 \text{ lb} \times \left(\dfrac{453.59 \text{ g}}{1 \text{ lb}}\right) \times \left(\dfrac{1 \text{ kg}}{1000 \text{ g}}\right) = 0.27 \text{ kg}$

(b) $0.60 \text{ lb} \times \left(\dfrac{453.59 \text{ g}}{1 \text{ lb}}\right) = 2.7 \times 10^2 \text{ g}$

(c) $0.60 \text{ lb} \times \left(\dfrac{453.59 \text{ g}}{1 \text{ lb}}\right) \times \left(\dfrac{1000 \text{ mg}}{1 \text{ g}}\right) = 2.7 \times 10^5 \text{ mg}$

1.31 $1.0 \text{ L} = 1000 \text{ mL} = 1000 \text{ cm}^3$

4 Chapter 1

$$\sqrt[3]{1000} = 10 \text{ cm} \times 10 \text{ cm} \times 10 \text{ cm}$$

Therefore, each side of a cube whose volume is 1.0 L is 10 cm.

1.33 $17 \text{ mm} \times \left(\dfrac{1 \text{ cm}}{10 \text{ mm}}\right) = 1.7 \text{ cm}$

Volume = L × W × H = 1.7 cm × 6.5 cm × 3.25 cm = 35.9 cm^3

Mass = Density × Volume = $2.2 \dfrac{\text{g}}{\text{cm}^3} \times 35.9 \text{ cm}^3 = 79 \text{ g}$

1.35 $°C = (°F - 32°F)\left(\dfrac{5°C}{9°F}\right)$

$°C = (68°F - 32°F)\left(\dfrac{5°C}{9°F}\right) = 20°C$

1.37 K = °C + 273 = 27°C + 273 = 300 K

1.39 $C_p = \left(\dfrac{\text{joules}}{\text{g} \times °\text{C}}\right) = \left(\dfrac{22.0 \text{ J}}{32 \text{ g} \times (28°C - 18°C)}\right) = 0.069 \dfrac{\text{J}}{\text{g} \times °\text{C}}$

1.41 $\rho = \text{density} = \left(\dfrac{\text{mass of solution}}{\text{volume of solution}}\right)$

Volume of solution = 19.84 mL

Mass of solution = mass of flask and solution − mass of flask

22.419 g = 54.381 g − 31.962 g

$\rho = \text{density} = \left(\dfrac{22.419 \text{ g}}{19.84 \text{ mL}}\right) = 1.130 \dfrac{\text{g}}{\text{mL}}$

1.43 Use the unit-conversion method.

$8.4 \times 10^3 \text{ kJ} \times \left(\dfrac{1 \text{ g}}{16.74 \text{ kJ}}\right) = 501 \text{ g}$

$501 \text{ g} \times \left(\dfrac{1 \text{ lb}}{453.6 \text{ g}}\right) = 1.1 \text{ lb}$

1.45 $\text{Density} = \left(\dfrac{\text{mass}}{\text{volume}}\right) = \left(\dfrac{41.242 - 28.463 \text{ g}}{12.3 \text{ mL}}\right) = \left(\dfrac{12.78 \text{ g}}{12.3 \text{ mL}}\right) = 1.04 \text{ g/mL}$

1.47 $\rho = \text{density} = 19.32 \dfrac{\text{g}}{\text{cm}^3}$

$2.416 \text{ kg} = 2416 \text{ g} \times \left(\dfrac{1 \text{ cm}^3}{19.32 \text{ g}}\right) = 125 \text{ cm}^3$

1.49 (a) Water changes into hydrogen gas and loses its identity; therefore, this is a chemical property.

(b) Lead does not change identity; therefore, this is a physical property.

(c) These properties of lead do not affect its identity; therefore, this is a physical property.

(d) Lead changes its identity and its properties; therefore, this is a chemical property.

1.51 (a) $9{,}620{,}000 \text{ kg} = 9.62 \times 10^6 \text{ kg}$

(b) $54{,}870 \text{ days} = 5.487 \times 10^4 \text{ days}$

(c) $253 \text{ milliseconds} \times \left(\dfrac{1 \text{ s}}{1000 \text{ ms}}\right) = 0.253 \text{ s} = 2.53 \times 10^{-1} \text{ s}$

(d) $0.000274 \text{ kilometers} \times \left(\dfrac{1000 \text{ m}}{1 \text{ km}}\right) = 2.74 \times 10^{-4} \text{ m}$

1.53 $45.8 \text{ mg} \times \left(\dfrac{1 \text{ oz}}{28350 \text{ mg}}\right) = 1.62 \times 10^{-3} \text{ oz}$

1.55 (a) $(8.2 \times 10^2) + (3.75 \times 10^4) = 820 + 37{,}500 = 3.83 \times 10^4$

(b) $(5.21 \times 10^{-2}) + (2.74 \times 10^{-3}) = 0.0521 + 0.00274 = 5.48 \times 10^{-2}$

(c) $(1.01 \times 10^{-4}) + (7.23 \times 10^{-3}) = 0.000101 + 0.00723 = 7.33 \times 10^{-3}$

1.57 $1.0 \text{ L} \times \left(\dfrac{1000 \text{ mL}}{1 \text{ L}}\right) = 1000 \text{ mL} = 1 \times 10^3 \text{ cm}^3$

Chapter 1

1.59 Heat lost by man = heat gained by water

Mass of water = 1000 gallons $\times \left(\dfrac{3.7853 \text{ L}}{1 \text{ gal}} \right) \times \left(\dfrac{1000 \text{ mL}}{1 \text{ L}} \right) \times \left(\dfrac{1 \text{ g}}{1 \text{ mL}} \right)$ = 3,785,300 g

= $C_p \times$ mass $\times \Delta T$ = 4.184 J/g × °C × 3,785,300 g × 2.49 Celsius degrees = 3.94×10^7 J in 1.5 h

$\dfrac{3.94 \times 10^7 \text{ J}}{1.5 \text{ hours}} \times 24$ hours = 6.3×10^8 J per day

1.61 The unspecified solution contains two components, NaCl and water. You could obtain pure NaCl by evaporating all the water from the mixture. Five grams of pure NaCl could then be weighed out in a laboratory, and the specified mixture of salt and water could be easily prepared.

1.63 You must modify your original hypothesis and retest it.

Chapter 2

Atomic Structure

2.1 The conservation of mass states that matter cannot be created or destroyed. No mass is lost as a result of a chemical reaction.

2.3 $\dfrac{1.520 \text{ g C}}{3.800 \text{ g}} = 4.000 \times 10^{-1} \times 100 = 40.00\%$

$\dfrac{0.2535 \text{ g H}}{3.800 \text{ g}} = 6.671 \times 10^{-2} \times 100 = 6.671\%$

$\dfrac{2.027 \text{ g O}}{3.800 \text{ g}} = 5.334 \times 10^{-1} \times 100 = 53.33\%$

2.5 Atomic mass of O = $\dfrac{4}{3}(12.0000) = 16.0000$ amu

2.7 Atomic mass of N = 0.875×16.0000 amu = 14.0 amu

2.9 Proton: mass 1.00728 amu, charge = 1+

Neutron: mass 1.00894 amu, charge = 0

Electron: mass 0.0005414 amu, charge = 1−

2.11 Atomic mass is the sum of the protons and the neutrons in the atom. In most atoms, the number of protons is close to the number of neutrons and therefore the atomic mass will be approximately double the number of protons, which equals the atomic number of an atom.

2.13 Because the number of electrons is different from the number of protons, it would be a charged ion. Because there are three more protons than electrons, it would be a cation with a charge of 3+.

2.15 Atomic mass of Sr:

$\dfrac{0.56}{100}(83.91) + \dfrac{9.86}{100}(85.91) + \dfrac{7.02}{100}(86.91) + \dfrac{82.56}{100}(87.91) = 87.62$ amu

2.17 $^{16}_{8}\text{O}, ^{17}_{8}\text{O}; ^{24}_{12}\text{Mg}, ^{25}_{12}\text{Mg}; ^{28}_{14}\text{Si}, ^{29}_{14}\text{Si}$

2.19 (a) Lithium; (b) Calcium; (c) Sodium; (d) Phosphorus; (e) Chlorine

Chapter 2

2.21 Inspection of the periodic table shows that:

Element	Group	Period
Li	I	2
Na	I	3
K	I	4
Rb	I	5
Cs	I	6

2.23 Li, Na, K, Rb and Cs are metals; they are all on the left side of the periodic table.

2.25 Main-group elements

2.27 Group VII

2.29 Emission occurs when an atom loses energy in going from a high-energy state to a lower one. The ground state is the lowest energy state and therefore the atom cannot undergo a transition to a lower state.

2.31 Shells are identified by a principal quantum number, n. Subshells within shells are identified by the letters s, p, d, f, etc. Orbitals are within subshells.

2.33 An orbital defines the probability of finding an electron in a region of space around an atomic nucleus.

2.35 There is one orbital in an s subshell.

2.37 No, an s subshell is spherically symmetrical.

2.39 The number of electrons must equal the number of protons for an element in its standard state. The number of protons identifies the element.

(a) Number of electrons = 12 = number of protons. The element with 12 protons is magnesium.

(b) Number of electrons = 16 = number of protons. The element with 16 protons is sulfur.

(c) Number of electrons = 17 = number of protons. The element with 17 protons is potassium.

2.41 The positive charge indicates that there is one more proton than electron. $1s^2 2s^2 2p^6$ = 10 electrons. Because there must be 11 protons, the cation is Na^+.

2.43 Group I. All elements possess one outer electron, ns^1.

2.45 Group II. All elements possess two outer electrons, ns^2.

2.47 True

2.49 Its energy increases.

2.51 They have virtually identical chemical properties owing to similar outer-shell electronic structure.

2.53 Calcium is a metal; it is found on the left side of the periodic table.

2.55 $^{16}_{8}O$, $^{18}_{8}O$

2.57 $93 \text{ e}^- \times \dfrac{1 \text{ amu}}{1835 \text{ e}^-} = 0.05 \text{ amu}$

$\dfrac{0.05 \text{ amu}}{237 \text{ amu}} = 0.022\%$

2.59 Because energy is proportional to frequency, blue light will have the greater energy.

2.61 The octet rule

2.63 One needs to know the natural abundances of each of the isotopes.

2.65 Dalton's atom was indestructible, but the "modern" atom can be decomposed into subatomic particles. In addition, the discovery of isotopes showed that the masses of the atoms of an element are not identical.

Chapter 3
Molecules and Chemical Bonds

3.1 Chemical-bond formation is the result of a process that allows the reacting elements to achieve stability by sharing or transferring electrons.

3.3 By losing, gaining, or sharing electrons.

3.5 These compounds are ionic; thus we do not need to use Greek prefixes (di, tri, etc.)

(a) K^+ = potassium ion; $(SO_4)^{2-}$ = sulfate ion → K_2SO_4 = potassium sulfate

(b) OH^- = hydroxide ion; Mn^{2+} = manganese (II) ion → $Mn(OH)_2$ = manganese (II) hydroxide

(c) NO_3^- = nitrate ion; Fe^{2+} = iron (II) ion → $Fe(NO_2)_2$ = iron (II) nitrate

(d) K^+ = potassium ion; $(H_2PO_4)^-$ = dihydrogen phosphate → KH_2PO_4 = potassium dihydrogen phosphate

(e) Ca^{2+} = calcium ion; $C_2H_3O_2^-$ = acetate ion → $Ca(C_2H_3O_2)_2$ = calcium acetate

(f) Na^+ = sodium ion; CO_3^{2-} = carbonate ion → Na_2CO_3 = sodium carbonate

3.7 (a) Ammonium; (b) nitrate; (c) sulfate; (d) phosphate; (e) acetate

3.9 (a) Aluminum = Al^{3+}; chloride = Cl^-. Compound must be neutral; therefore $AlCl_3$.

(b) Aluminum = Al^{3+}; Sulfide = S^{2-}. Compound must be neutral; therefore Al_2S_3.

(c) Aluminum = Al^{3+}; Nitride ion = N^{3-}. Compound must be neutral; therefore AlN.

3.11 (a) Lithium = Li^+; oxide = O^{2-}. Compound must be neutral; therefore Li_2O.

(b) Calcium = Ca^{2+}; bromide = Br^-. Compound must be neutral; therefore $CaBr_2$.

(c) Aluminum = Al^{3+}; oxide = O^{2-}. Compound must be neutral; therefore Al_2O_3.

(d) Sodium = Na^+; sulfide = S^{2-}. Compound must be neutral; therefore Na_2S.

3.13 Group I forms 1+ ions; Group VII forms 1– ions. For neutral compound, ratio = 1:1.

3.15 Group II forms 2+ ions; Group VII forms 1– ions. For neutral compound, ratio = 1:2.

3.17 Calcium ion = Ca^{2+}. Any Group VI element forms a 2– ion. Therefore, to make a neutral compound, one calcium and one Group VI element must be included in the compound. The first three elements in Group VI are oxygen, O; sulfur, S; and selenium, Se. Thus,

(a) CaO (b) CaS (c) CaSe

3.19 Aluminum ion = Al^{3+}. Any Group VI element forms a 2– ion. Therefore, to make a neutral compound, two aluminums and three Group VI elements must be included in the compound. The first three elements in Group VI are oxygen, O; sulfur, S; and selenium, Se. Thus,

(a) Al_2O_3 (b) Al_2S_3 (c) Al_2Se_3

3.21 Aluminum ion = Al^{3+}. Any Group VII element forms a 1– ion. Therefore, to make a neutral compound, one aluminum and three Group VII elements must be included in the compound. The first four elements in Group VII are fluorine, F; chlorine, Cl; bromine, Br; iodine, I. Thus,

(a) AlF_3 (b) $AlCl_3$ (c) $AlBr_3$ (d) AlI_3

3.23 Procedure to draw Lewis dot structure:

(i) Place atoms and connect bonds.

(ii) Determine how many valence electrons are available from all atoms

(iii) Subtract 2 electrons for each bond drawn in part i.

(iv) Distribute remaining electrons among the atoms. Result should be that every atom has 8 electrons associated with it either from free pairs of electrons or from covalent bonds.

CCl_4

(i)

```
         Cl
         |
   Cl — C — Cl
         |
         Cl
```

(ii) 4 electrons from the carbon + (7 electrons from each chlorine × 4) = 32 e^-.

(iii) Covalent bonds use 8 electrons. 32 – 8 = 24 electrons remain.

(iv) Distribute 24 electrons among four Cl atoms (6 each) to give each Cl 8 electrons.

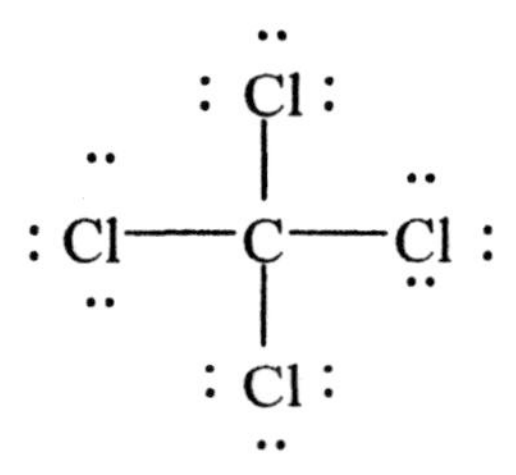

$$:\ddot{\underset{..}{Cl}}-\underset{:\ddot{Cl}:}{\overset{:\ddot{Cl}:}{\underset{|}{\overset{|}{C}}}}-\ddot{\underset{..}{Cl}}:$$

Check that each atom has 8 electrons associated with it. ---- ✓

3.25 Use steps i through iv in the solution to Exercise 3.23:

(i)

$$F-O-F$$

(ii) 6 electrons from the oxygen + (7 electrons from each fluorine × 2) = 20 e⁻.

(iii) Covalent bonds use 4 electrons. 20 – 4 = 16 electrons remain.

(iv) Distribute 16 electrons among F and O. First put 4 electrons (two free pairs) on the oxygen to give it an octet. That leaves 12 electrons; distribute these electrons among two F atoms (6 each) to give each F 8 electrons.

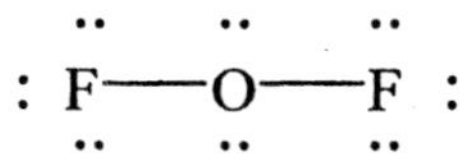

$$:\ddot{\underset{..}{F}}-\ddot{\underset{..}{O}}-\ddot{\underset{..}{F}}:$$

Check that each atom has 8 electrons associated with it. ---- ✓

3.27 Use steps i through iv in the solution to Exercise 3.23:

(i)

$$O-C-O$$

(ii) 4 electrons from the carbon + (6 electrons from each oxygen × 2) = 16 e⁻.

(iii) Covalent bonds use 4 electrons. 16 – 4 = 12 electrons remain.

(iv) Distribute 12 electrons among C and O. First put 4 electrons (two free pairs) on the carbon to give it an octet. That leaves 8 electrons, distribute these electrons among two O atoms (4 each) to give:

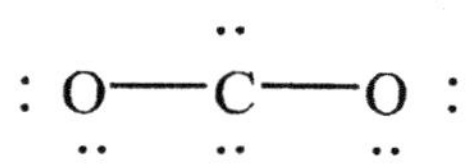

$$:\!\ddot{O}\!-\!\ddot{C}\!-\!\ddot{O}\!:$$

Check that each of the atoms has 8 electrons associated with it. ---- **NO**

Both oxygens have only six electrons. To remedy this, we need to incorporate multiple (double) bonds. When there are not enough electrons to achieve an octet for each atom, changing single bonds to double bonds will result in more sharing of electrons.

Try again:

(i) Make bonds between C and O double bonds:

$$O=C=O$$

(ii and iii) Covalent bonds use 8 electrons → 16 – 8 = 8 electrons remain.

(iv) Distribute remaining electrons between the two oxygens, because the carbon has 8 electrons associated with it.

$$:\!\ddot{O}\!=\!C\!=\!\ddot{O}\!:$$

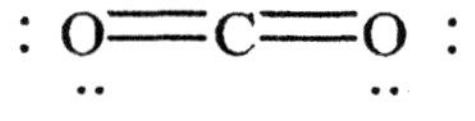

Check that each atom has 8 electrons associated with it. ---- ✓

3.29 VSEPR provides information on the three-dimensional structure of a molecule by taking into account the fact that electrons have a negative charge. Therefore each pair of electrons in an atom wants to be as far away from all other pairs of electrons as possible. In this theory, "pairs of electrons" means either nonbonded pairs of electrons or covalent bonds. Furthermore, single, double, and triple bonds each count as a single "pair of electrons."

Therefore, before using VSEPR to determine the three-dimensional arrangement of the bonds or nonbonded electron pairs, we must first determine the Lewis dot structures. We will use the procedure outlined in the solution to Exercise 3.23.

(i)

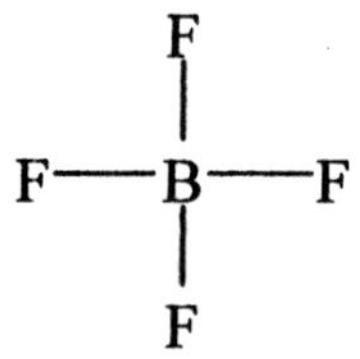

(ii) 3 electrons from the boron + (7 electrons from each fluorine × 4) + 1 extra electron from the (−) charge = 32 e⁻.

(iii) Covalent bonds use 8 electrons. 32 − 8 = 24 electrons remain.

(iv) Distribute 24 electrons among four fluorine atoms (6 each) to give:

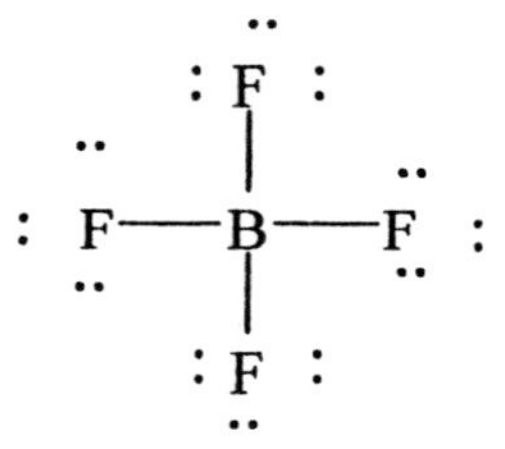

$$\begin{array}{c} ..\\ :F:\\ ..\ |\ ..\\ :F-B-F:\\ ..\ |\ ..\\ :F:\\ .. \end{array}$$

Therefore, the central atom is boron, and it has four bonds, which equals four "pairs of electrons" around it. These four bonds must arrange so that they are as far away from each other as possible. This is accomplished by the four fluorine atoms being situated at the corners of a tetrahedron, with the boron atom at the center.

3.31 The Lewis dot structures are taken from the solutions to Exercises 3.23, 3.29, and 3.27 for parts a, b, and c, respectively.

(a)

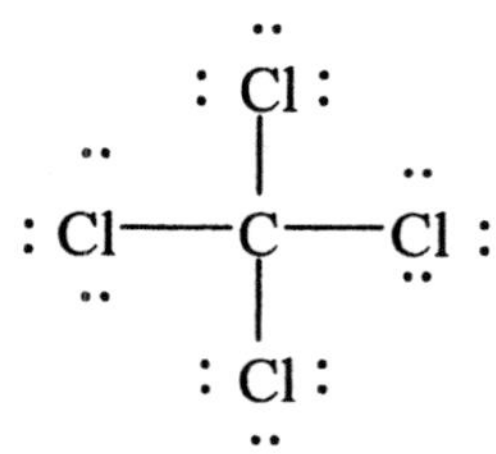

$$\begin{array}{c} ..\\ :Cl:\\ ..\ |\ ..\\ :Cl-C-Cl:\\ ..\ |\ ..\\ :Cl:\\ .. \end{array}$$

From this structure and VSEPR, CCl$_4$ forms a tetrahedron with each C–Cl bond going from the center to the corner. Even though each C–Cl bond is polar, this arrangement results in a cancellation of these polar contributions to give a molecule with no dipole moment.

(b)

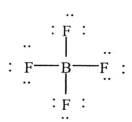

From this structure and VSEPR, BF₄ forms a tetrahedron with each B–F bond going from the center to the corner. Even though each B–F bond is polar, this arrangement results in a cancellation of these polar contributions to give a molecule with no dipole moment.

(c)

$$:\!\ddot{O}\!=\!C\!=\!\ddot{O}\!:$$

From this structure and VSEPR, CO_2 forms a linear molecule with each C=O 180° from each other. Even though each C=O bond is polar, this arrangement results in a cancellation of these polar contributions to give a molecule with no dipole moment.

3.33 A chemical bond that consists of two electrons with paired spins.

3.35 These compounds are ionic; thus we do not need to use Greek prefixes (di, tri, etc.).

(a) Ag^+ = silver ion; NO_3^- = nitrate ion → $AgNO_3$ = silver nitrate.

(b) Hg^{2+} = Mercury (II) ion; Cl^- = chloride ion → $HgCl_2$ = mercury (II) chloride.

(c) Na^+ = Sodium ion; CO_3^{2-} = carbonate ion → Na_2CO_3 = sodium carbonate.

(d) Ca^{2+} = calcium ion; PO_4^{3-} = phosphate → $Ca_3(PO_4)_2$ = calcium phosphate.

3.37 Use names of ions in the solution to Exercise 3.36.

(a) Magnesium chloride   (b) Magnesium sulfide   (c) Magnesium nitride

3.39 Ba^{2+} = barium ion; O^{2-} = oxide ion; S^{2-} = sulfide ion; Se^{2-} = selenide ion. Therefore:

(a) Barium oxide   (b) Barium sulfide   (c) Barium selenide

3.41 K^+ = Potassium ion; N^{3-} = nitride ion; P^{3-} = phosphide ion; As^{3-} = arsenide ion. Therefore:

(a) Potassium nitride   (b) Potassium phosphide   (c) Potassium arsenide

3.43 Te is in the same group as oxygen (Group VI) and Br is in the same group as F (Group VII). Therefore the Lewis dot structure of TeBr$_2$ will be very similar to that of OF$_2$, as determined in the solution to Exercise 3.25:

$$:\ddot{\underset{..}{Br}}\!-\!\ddot{Te}\!-\!\ddot{\underset{..}{Br}}:$$

The central atom is Te, and it has two bonds and two nonbonded pairs, which equals four "pairs of electrons" around it. These four "pairs" must arrange so that they are as far away from each other as possible. This is accomplished by the two bromine atoms and the two nonbonded pairs being situated at the corners of a tetrahedron. Thus, the Br–Te–Br bonds will form a bent molecule.

3.45 Ionic bonds form when the ion attraction power between two atoms is very different (see Figure 3.1). Ion attraction power is related to the difference between the ionization energies of the two elements.

Ionization energies follow the periodic table as shown in Figure 3.1. Therefore, if two atoms are close to each other in the periodic table, they will form covalent bonds; otherwise they will form ionic bonds.

(a) Si and O are close; therefore covalent bonds.

(b) Be and C are close; therefore covalent bonds.

(c) C and N are close; therefore covalent bonds.

(d) Ca and Br are not close; therefore ionic bonds.

(e) B and N are close; therefore covalent bonds.

3.47 Lithium ion = Li$^+$. Any Group VII element forms a 1– ion. Therefore, to make a neutral compound, one lithium and one group VII element must be included in the compound. The first 4 elements in Group VII are fluorine, F; chlorine, Cl; bromine, Br; iodine, I. Thus,

(a) LiF   (b) LiCl   (c) LiBr   (d) LiI

3.49 Yes. Hydrogen requires only two electrons. Also, beryllium and boron require four and six electrons, respectively, to complete their valence shells.

3.51 Test its ability to conduct electricity when dissolved in water. An ionic substance will dissociate into ions that will conduct electricity. Thus, if the dissolved substance conducts electricity, it is ionic.

Molecules and Chemical Bonds   17

3.53 Use steps i through iv in the solution to Exercise 3.23:

(i)
$$H\text{—}O\text{—}O\text{—}H$$

(ii) 2 electrons from the hydrogens + (6 electrons from each oxygen × 2) = 14 e⁻.

(iii) Covalent bonds use 6 electrons. 14 – 6 = 8 electrons remain.

(iv) Distribute the 8 electrons to the oxygen atoms to give:

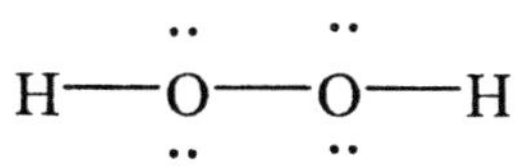

$$H\text{—}\ddot{\underset{..}{O}}\text{—}\ddot{\underset{..}{O}}\text{—}H$$

Check that each atom has 8 electrons associated with it. ---- ✓

3.55 Use steps i through iv in the solution to Exercise 3.23:

(i)

$$\begin{array}{c} O \qquad\qquad O \\ \diagdown \qquad\qquad \diagup \\ N\text{—}N \\ \diagup \qquad\qquad \diagdown \\ O \qquad\qquad O \end{array}$$

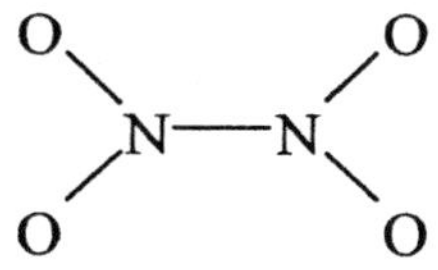

(ii) (5 electrons from each nitrogen × 2) + (6 electrons from each oxygen × 4) = 34 e⁻.

(iii) Covalent bonds use 10 electrons. 34 – 10 = 24 electrons remain.

(iv) Distributing these 24 electrons among the nitrogen and the oxygen atoms to fill the octets does not work. Therefore, as in the solution to Exercise 3.27, make some of the bonds double bonds. When two N–O bonds are changed to double bonds, the correct resultant Lewis dot structure is:

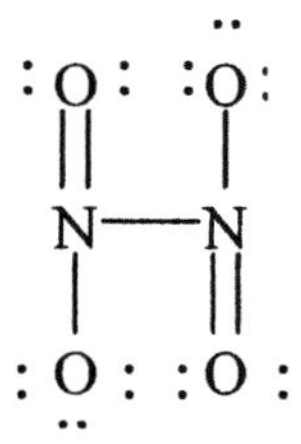

$$\begin{array}{c} :\ddot{O}:\ \ :\ddot{O}: \\ \|\quad\ \ | \\ N\text{—}N \\ |\quad\ \ \| \\ :\ddot{O}:\ \ :\ddot{O}: \end{array}$$

Check that each atom has 8 electrons associated with it. ---- ✓

Chapter 4

Chemical Calculations

4.1 To determine the formula masses of a compound, add the formula masses of all the atoms:

(a) CaCrO$_4$	Atom	Mass	Total
	Ca	40.1	40.1
	Cr	52.0	52.0
	O	16.0	4 × 16.0
		Total:	156.1

(b) Mg(OH)$_2$	Atom	Mass	Total
	Mg	24.3	24.3
	H	1.01	2 × 1.01
	O	16.0	2 × 16.0
		Total:	58.32

(c) TiCl$_4$	Atom	Mass	Total
	Ti	47.88	47.88
	Cl	35.45	4 × 35.45
		Total:	189.7

(d) Na$_2$Cr$_2$O$_7$	Atom	Mass	Total
	Na	22.99	2 × 22.99
	Cr	52.0	2 × 52.0
	O	16.0	7 × 16.0
		Total:	262.0

(e) C$_3$H$_7$OH	Atom	Mass	Total
	C	12.01	3 × 12.01
	H	1.01	8 × 1.01
	O	16.0	1 × 16.0
		Total:	60.11

Chemical Calculations

4.3 To determine the formula masses of a compound, add the formula masses of all the atoms:

(a) $Zn_3(AsO_4)_2$

Atom	Mass	Total
Zn	65.39	3 × 65.39
As	74.92	2 × 74.92
O	16.0	8 × 16.0
Total:		474.0

(b) $Al(C_2H_3O_2)_3$

Atom	Mass	Total
Al	26.98	26.98
C	12.01	6 × 12.01
H	1.01	9 × 1.01
O	16.0	6 × 16.0
Total:		204.1

(c) $HgCl_2$

Atom	Mass	Total
Hg	200.59	1 × 200.59
Cl	35.45	2 × 35.45
Total:		271.5

(d) $Sr(ClO_3)_2$

Atom	Mass	Total
Sr	87.62	87.62
Cl	35.45	2 × 35.45
O	16.0	6 × 16.0
Total:		254.5

(e) BI_3

Atom	Mass	Total
B	10.81	1 × 10.81
I	126.9	3 × 126.9
Total:		391.5

4.5 To convert from the number of grams into moles, the number of grams must be divided by the formula weight:

$$24.67 \text{ g NaCl} \times \frac{1 \text{ mol NaCl}}{58.43 \text{ g NaCl}} = 0.4221 \text{ mol NaCl}$$

Chapter 4

4.7 To convert the number of moles into grams, the number of moles must be multiplied by the formula weight:

$$0.275 \text{ mol Na}_5\text{P}_3\text{O}_{10} \times \frac{367.86 \text{ g Na}_5\text{P}_3\text{O}_{10}}{1 \text{ mol Na}_5\text{P}_3\text{O}_{10}} = 101 \text{ g Na}_5\text{P}_3\text{O}_{10}$$

4.9 To convert moles into the number of molecules, the number of moles must be multiplied by Avogadro's number, 6.022×10^{23} molecules/mol. In this problem, one must take into account the fact that each molecule creates two Zn ions.

$$30.5 \text{ g Zn}_2\text{P}_2\text{O}_7 \times \frac{1 \text{ mol}}{304.7 \text{ g}} \times \frac{2 \text{ mol Zn ions}}{1 \text{ mol Zn}_2\text{P}_2\text{O}_7} \times \frac{6.022 \times 10^{23} \text{ ion}}{1 \text{ mol Zn ions}} = 1.20 \times 10^{23} \text{ Zn ions}$$

4.11 To convert the number of molecules into moles, the number of molecules must be divided by Avogadro's number, 6.022×10^{23} molecules/mol.

$$10.54 \times 10^{24} \text{ molecules} \times \frac{1 \text{ mol}}{6.022 \times 10^{23} \text{ molecules}} \times \frac{16.05 \text{ g CH}_4}{1 \text{ mol CH}_4} = 280.8 \text{ g CH}_4$$

4.13 Apply these proportions to a sample that is 100 grams:

$$36.05 \text{ g Zn} \times \frac{1 \text{ mol}}{65.39 \text{ g}} = 0.5513 \text{ mol Zn}$$

$$28.67 \text{ g Cr} \times \frac{1 \text{ mol}}{52.8 \text{ g}} = 0.5513 \text{ mol Cr}$$

$$35.28 \text{ g O} \times \frac{1 \text{ mol}}{16 \text{ g}} = 2.2 \text{ mol O}$$

Divide each molar amount by the smallest value to get whole numbers:

$0.5513/0.5513 = 1$ $\qquad$ $0.5513/0.5513 = 1$ $\qquad$ $2.2/0.5513 = 3.99 \approx 4$

Thus, the empirical formula $ZnCrO_4$

4.15 (a) To balance the equation, concentrate on one atom at a time.

(i) Because there are two N atoms on the left-hand side of the equation and only one N atom on the right-hand side of the equation, put a coefficient of 2 in front of NBr_3 to equate the number of N atoms on both sides of the reaction equation:

$$N_2 + Br_2 \rightarrow 2 \text{ NBr}_3$$

(ii) Now because there are six Br atoms on the right-hand side and only two Br atoms on the left-hand side, put a coefficient of 3 in front of Br_2 to equate the number of Br atoms on both sides of the reaction equation:

$$N_2 + 3\ Br_2 \rightarrow 2\ NBr_3$$

(b) $2\ HNO_3 + Ba(OH)_2 \rightarrow Ba(NO_3)_2 + 2\ H_2O$

(c) $HgCl_2 + H_2S \rightarrow HgS + 2\ HCl$

4.17 (a) $4\ P + 5\ O_2 \rightarrow 2\ P_2O_5$

(b) $FeCl_2 + K_2SO_4 \rightarrow FeSO_4 + 2\ KCl$

(c) $HgCl + 2\ NaOH + 2\ NH_4Cl\ (\ Hg(NH_3)_2Cl + 2\ NaCl + 2\ H_2O$

4.19 Hydrocarbons combust to form CO_2 and H_2O:

$$C_6H_6 + O_2 \rightarrow CO_2 + H_2O$$

Now balance as in the solutions to Exercises 4.15 through 4.18.

(i) Because there are six C atoms on the left-hand side of the equation and only one C atom on the right-hand side of the equation, put a coefficient of 6 in front of CO_2 to equate the number of C atoms on both sides of the reaction equation:

$$C_6H_6 + O_2 \rightarrow 6\ CO_2 + H_2O$$

(ii) Now because there are six H atoms on the left-hand side and only two H atoms on the right-hand side, put a coefficient of 3 in front of H2O to equate the number of H atoms on both sides of the reaction equation:

$$C_6H_6 + O_2 \rightarrow 6\ CO_2 + 3\ H_2O$$

(iii) Now there are two O atoms on the left-hand side and 15 O atoms on the right-hand side; the simplest way to equate the number of O atoms on both sides of the equation is to place a stoichiometric prefactor of 15/2 in front of O_2:

$$C_6H_6 + 15/2\ O_2 \rightarrow 6\ CO_2 + 3\ H_2O$$

(iv) Multiply through by 2 to obtain integer coefficients:

$$2\ C_6H_6 + 15\ O_2 \rightarrow 12\ CO_2 + 6\ H_2O$$

Chapter 4

4.21 $C_5H_{12} + 8\ O_2 \rightarrow 5\ CO_2 + 6\ H_2O$

$2\ C_8H_{18} + 25\ O_2 \rightarrow 16\ CO_2 + 18\ H_2O$

$C_{10}H_{20} + 15\ O_2 \rightarrow 10\ CO_2 + 10\ H_2O$

4.23 $CH_4 + 2\ O_2 \rightarrow CO_2 + 2\ H_2O$

4.25 Unit-conversion factors are ratios of amount of one component to amount of another component, in moles:

$$\frac{2\ Al(OH)_3}{3\ H_2SO_4}; \frac{1\ Al(OH)_3}{3\ H_2O}; \frac{1\ H_2SO_4}{2\ H_2O}; \frac{2\ Al(OH)_3}{1\ Al_2(SO_4)_3}; \frac{3\ H_2SO_4}{1\ Al_2(SO_4)_3}; \frac{1\ Al_2(SO_4)_3}{6\ H_2O}$$

and the inverse of each of them.

4.27 From Exercise 4.25:

$$2\ Al(OH)_3 + 3\ H_2SO_4 \rightarrow Al_2(SO_4)_3 + 6\ H_2O$$

$$2.5\ \text{mol}\ H_2SO_4 \times \frac{1\ \text{mol}\ Al_2(SO_4)_3}{3\ \text{mol}\ H_2SO_4} = 0.83\ \text{mol}\ Al_2(SO_4)_3$$

4.29 $6.5\ \text{mol}\ H_2O \times \dfrac{3\ \text{mol}\ H_2SO_4}{6\ \text{mol}\ H_2O} = 3.3\ \text{mol}\ H_2SO_4$

4.31 Note: Information is in grams, answer is in grams; must still do conversions in moles:

$$46.0\ \text{g}\ H_2SO_4 \times \frac{1\ \text{mol}\ H_2SO_4}{98\ \text{g}\ H_2SO_4} = 0.47\ \text{mol}\ H_2SO_4$$

$$0.47\ \text{mol}\ H_2SO_4 \times \frac{1\ \text{mol}\ Al_2(SO_4)_3}{3\ \text{mol}\ H_2SO_4} = 0.156\ \text{mol}\ Al_2(SO_4)_3$$

$$0.156\ \text{mol}\ Al_2(SO_4)_3 \times \frac{246\ \text{g}\ Al_2(SO_4)_3}{1\ \text{mol}\ Al_2(SO_4)_3} = 38.47\ \text{g}\ Al_2(SO_4)_3$$

4.33 $2\ Cu(NO_3)_2 + 4\ NaI \rightarrow I_2 + 2\ CuI + 4\ NaNO_3$

(a) $0.87\ \text{mol}\ NaI \times \dfrac{2\ \text{mol}\ Cu(NO_3)_2}{4\ \text{mol}\ NaI} = 0.44\ \text{mol}\ Cu(NO_3)_2$

(b) $1.45\ \text{mol}\ CuI \times \dfrac{4\ \text{mol}\ NaI}{2\ \text{mol}\ CuI} = 2.91\ \text{mol}\ NaI$

(c) $15.0 \text{ g NaI} \times \dfrac{1 \text{ mol NaI}}{149.89 \text{ g NaI}} = 0.1 \text{ mol NaI}$

$0.1 \text{ mol NaI} \times \dfrac{2 \text{ mol Cu(NO}_3)_2}{4 \text{ mol NaI}} = 0.05 \text{ mol Cu(NO}_3)_2$

$0.05 \text{ mol Cu(NO}_3)_2 \times \dfrac{187.55 \text{ g Cu(NO}_3)_2}{1 \text{ mol Cu(NO}_3)_2} = 9.38 \text{ g Cu(NO}_3)_2$

(d) $15.00 \text{ g NaI} \times \dfrac{1 \text{ mol NaI}}{149.89 \text{ g NaI}} = 0.1 \text{ mol NaI}$

$0.1 \text{ mol NaI} \times \dfrac{4 \text{ mol NaNO}_3}{4 \text{ mol NaI}} = 0.1 \text{ mol NaNO}_3$

$0.1 \text{ mol NaNO}_3 \times \dfrac{84.99 \text{ g NaNO}_3}{1 \text{ mol NaNO}_3} = 8.49 \text{ g NaNO}_3$

4.35 (a) $Ca_3(PO_4)_2 + 4 H_3PO_4 \rightarrow 3 Ca(H_2PO_4)_2$

(b) $FeCl_2 + (NH_4)_2S \rightarrow FeS + 2 NH_4Cl$

(c) $2 KClO_3 \rightarrow 2 KCl + 3 O_2$

(d) $3 O_2 \rightarrow 2 O_3$

(e) $2 C_6H_6 + 15 O_2 \rightarrow 12 CO_2 + 6 H_2O$ ← wrong

4.37 (a) $102.6 \text{ g BaCO}_3 \times \dfrac{1 \text{ mol BaCO}_3}{197 \text{ g BaCO}_3} = 0.5199 \text{ mol BaCO}_3$

(b) $60.75 \text{ g HBr} \times \dfrac{1 \text{ mol HBr}}{80.9 \text{ g HBr}} = 0.7508 \text{ mol HBr}$

(c) $148.5 \text{ g CuCl} \times \dfrac{1 \text{ mol CuCl}}{99 \text{ g CuCl}} = 1.500 \text{ mol CuCl}$

(d) $50.4 \text{ g HNO}_3 \times \dfrac{1 \text{ mol HNO}_3}{63 \text{ g HNO}_3} = 0.800 \text{ mol HNO}_3$

(e) $65 \text{ g C}_6H_{12}O_6 \times \dfrac{1 \text{ mol C}_6H_{12}O_6}{180.5 \text{ g C}_6H_{12}O_6} = 0.36 \text{ mol C}_6H_{12}O_6$

Chapter 4

4.39 The formula weight for $Ca(OH)_2$ is 74.07 grams/mole.

$$133 \text{ g Ca(OH)}_2 \times \frac{1 \text{ formula unit Ca(OH)}_2}{74.07 \text{ formula weight Ca(OH)}_2} = 1.80 \text{ formula units Ca(OH)}_2$$

4.41 When sodium is oxidized, each atom loses one electron:

$$Na \rightarrow Na^+ + 1 \text{ e}^-$$

$$42 \text{ g Na} \times \frac{1 \text{ mol Na}}{22.98 \text{ g Na}} \times \frac{6.02 \times 10^{23} \text{ Na atoms}}{1 \text{ mol Na}} \times \frac{1 \text{ e}^-}{1 \text{ Na atom}} = 1.1 \times 10^{24} \text{ electrons}$$

4.43 Assume 100 g sample of silver chloride:

$$75.26 \text{ g Ag} \times \frac{1 \text{ mol Ag}}{107.86 \text{ g Ag}} = 0.697 \text{ mol Ag} \qquad 24.74 \text{ g Cl} \times \frac{1 \text{ mol Cl}}{35.45 \text{ g Cl}} = 0.697 \text{ mol Cl}$$

Divide by the smaller number of relative moles: for both Ag and Cl, 0.697/0.697 = 1. Therefore, the empirical formula is AgCl.

4.45 Assume 100 g sample of the methyl ether:

$$52.17 \text{ g C} \times \frac{1 \text{ mol C}}{12.0 \text{ g C}} = 4.34 \text{ mol C} \qquad 13.05 \text{ g H} \times \frac{1 \text{ mol H}}{1.0 \text{ g H}} = 13.05 \text{ mol H}$$

$$34.78 \text{ g O} \times \frac{1 \text{ mol O}}{16.0 \text{ g O}} = 2.17 \text{ mol O}$$

Divide by the smaller number of relative moles: for C, 4.34/2.17 = 2; for H, 13.05/2.17 = 6, for O, 2.17/2.17 = 1. Therefore, the empirical formula is C_2H_6O.

4.47 (a) $AgNO_3 + KCl \rightarrow AgCl + KNO_3$

(b) $Ba(NO_3)_2 + Na_2SO_4 \rightarrow BaSO_4 + 2 \text{ NaNO}_3$

(c) $2 \text{ (NH}_4)_3PO_4 + 3 \text{ Ca(NO}_3)_2 \rightarrow Ca_3(PO_4)_2 + 6 \text{ NH}_4NO_3$

(d) $3 \text{ Mg(OH)}_2 + 2 \text{ H}_3PO_4 \rightarrow Mg_3(PO_4)_2 + 6 \text{ H}_2O$

4.49 Note: Correct compound is $Al_2(SO_4)_3$

$$2.000 \text{ mol Al}_2(SO_4)_3 \times \frac{342.15 \text{ g Al}_2(SO_4)_3}{1 \text{ mol Al}_2(SO_4)_3} = 684.3 \text{ g Al}_2(SO_4)_3$$

Chemical Calculations 25

4.51 Assume 1 mole of compound:

$$17 \text{ mol C} \times \frac{12.01 \text{ g C}}{1 \text{ mol C}} = 204.2 \text{ g C} \qquad \%\text{C}: \frac{204.2 \text{ g C}}{303.4 \text{ g total}} = 67.33\% \text{ C}$$

$$21 \text{ mol H} \times \frac{1.01 \text{ g H}}{1 \text{ mol H}} = 21.2 \text{ g H} \qquad \%\text{H}: \frac{21.2 \text{ g H}}{303.4 \text{ g total}} = 6.931\% \text{ H}$$

$$1 \text{ mol N} \times \frac{14.0 \text{ g N}}{1 \text{ mol N}} = 14.0 \text{ g N} \qquad \%\text{N}: \frac{14.0 \text{ g N}}{303.4 \text{ g total}} = 4.620\% \text{ N}$$

$$4 \text{ mol O} \times \frac{16.0 \text{ g O}}{1 \text{ mol O}} = 64 \text{ g O} \qquad \%\text{O}: \frac{64 \text{ g O}}{303.4 \text{ g total}} = 21.12\% \text{ O}$$

4.53 (a) $0.46 \text{ mol H}_2\text{SO}_4 \times \dfrac{97.8 \text{ g H}_2\text{SO}_4}{1 \text{ mol H}_2\text{SO}_4} = 45 \text{ g H}_2\text{SO}_4$

(b) $1.80 \text{ mol Ca(OH)}_2 \times \dfrac{74.1 \text{ g Ca(OH)}_2}{1 \text{ mol Ca(OH)}_2} = 133 \text{ g Ca(OH)}_2$

(c) $0.76 \text{ mol C}_2\text{H}_6\text{O} \times \dfrac{46.05 \text{ g C}_2\text{H}_6\text{O}}{1 \text{ mol C}_2\text{H}_6\text{O}} = 35 \text{ g C}_2\text{H}_6\text{O}$

(d) $1.89 \text{ mol C}_4\text{H}_{10} \times \dfrac{58.2 \text{ g C}_4\text{H}_{10}}{1 \text{ mol C}_4\text{H}_{10}} = 110 \text{ g C}_4\text{H}_{10}$

(e) $1.66 \text{ mol FeCl}_2 \times \dfrac{126.5 \text{ g FeCl}_2}{1 \text{ mol FeCl}_2} = 210 \text{ g FeCl}_2$

4.55 Assume 100 g sample of the oxide of phosphorus:

$$43.64 \text{ g P} \times \frac{1 \text{ mol P}}{30.97 \text{ g P}} = 1.41 \text{ mol P} \qquad 56.36 \text{ g O} \times \frac{1 \text{ mol O}}{16.0 \text{ g O}} = 3.52 \text{ mol O}$$

Divide by the smaller number of relative moles: for P, $1.41/1.41 = 1$; for O, $3.52/1.41 = 2.5$. Not all are integers; therefore, they must be multiplied by a factor (2) to convert them into integers to give the empirical formula P_2O_5.

The molecular mass was shown to be 282 g/mol, which is two times the empirical weight of 141 g/mol. Thus, the molecular formula is twice that calculated from the weight analysis: P_4O_{10}.

Chapter 5

The Physical Properties of Gases

5.1   (a) $364 \text{ mm Hg} \times \dfrac{1 \text{ torr}}{1 \text{ mm Hg}} = 364 \text{ torr}$

    (b) $483 \text{ mm Hg} \times \dfrac{1 \text{ torr}}{1 \text{ mm Hg}} = 483 \text{ torr}$

    (c) $675 \text{ mm Hg} \times \dfrac{1 \text{ torr}}{1 \text{ mm Hg}} = 675 \text{ torr}$

    (d) $735 \text{ mm Hg} \times \dfrac{1 \text{ torr}}{1 \text{ mm Hg}} = 735 \text{ torr}$

5.3   (a) $735 \text{ mm Hg} \times \dfrac{1 \text{ torr}}{1 \text{ mm Hg}} = 735 \text{ torr}$

    (b) $735 \text{ mm Hg} \times \dfrac{1 \text{ atm}}{760 \text{ mm Hg}} = 0.967 \text{ atm}$

5.5   (a) $1.20 \text{ atm} \times \dfrac{760 \text{ torr}}{1 \text{ atm}} = 912 \text{ torr}$

    (b) $0.450 \text{ atm} \times \dfrac{760 \text{ mm Hg}}{1 \text{ atm}} = 342 \text{ mm Hg}$

    (c) $850 \text{ mm Hg} \times \dfrac{1 \text{ atm}}{760 \text{ mm Hg}} = 1.12 \text{ atm}$

5.7   Temperature, T, and the number of moles, n.

5.9   Volume, V, and the number of moles, n.

5.11 $P_1V_1 = P_2V_2$

     $(1.0 \text{ atm})(2.0 \text{ L}) = (0.80 \text{ atm})(V_2)$ → $V_2 = 2.5 \text{ L}$

5.13 $\dfrac{P_1V_1}{T_1} = \dfrac{P_2V_2}{T_2}$ → $\dfrac{(1.0 \text{ L})(0.800 \text{ atm})}{298 \text{ K}} = \dfrac{(1.00 \text{ atm})V_2}{373 \text{ K}}$

The Physical Properties of Gases 27

$V_2 = 1.00$ L

5.15 $\dfrac{P_1V_1}{T_1} = \dfrac{P_2V_2}{T_2} \rightarrow \dfrac{(1.20\text{ L})(0.800\text{ atm})}{298\text{ K}} = \dfrac{(1.00\text{ atm})(2.00\text{ L})}{T_2}$

$T_2 = 621$ K $= 348°$C

5.17 $\dfrac{P_1V_1}{T_1} = \dfrac{P_2V_2}{T_2} \rightarrow \dfrac{(1.50\text{ atm})(1.20\text{ L})}{298\text{ K}} = \dfrac{P_2(0.600\text{ L})}{323\text{ K}}$

$P_2 = 3.25$ atm

5.19

$\dfrac{P_1V_1}{T_1} = \dfrac{P_2V_2}{T_2} \rightarrow P_2 = 2P_1;\ T_2 = 1.5T_1$

$\dfrac{P_1V_1}{T_1} = \dfrac{2P_1V_2}{1.5T_1} \rightarrow V_2 = \dfrac{P_1}{2P_1}\dfrac{1.5T_1}{T_1}V_1$

$V_2 = 0.75 \times V_1 = 2.25$ L

5.21 At STP, P = 1.0 atm and T = 273 K.

$\dfrac{P_1V_1}{T_1} = \dfrac{P_2V_2}{T_2} \rightarrow \dfrac{(0.830\text{ atm})(2.00\text{ L})}{298\text{ K}} = \dfrac{(1.00\text{ atm})V_2}{273\text{ K}}$

$V_2 = 1.52$ L

5.23 At STP, P = 1.00 atm; T = 273 K; 1 mole of gas occupies 22.4 L

$P = 1.00\text{ atm} \times \dfrac{760\text{ torr}}{1\text{ atm}} = 760$ torr; $T = 273$ K; $n = 1$ mole;

$V = 22.4\text{ L} \times \dfrac{1000\text{ mL}}{\text{L}} = 2.25 \times 10^4$ mL

$R = \dfrac{PV}{nT} = \dfrac{(760\text{ torr})(2.24 \times 10^4\text{ mL})}{(1\text{ mole})(273\text{ K})} = 6.24 \times 10^4\ \dfrac{\text{torr} \cdot \text{mL}}{\text{mole} \cdot \text{K}}$

5.25 $PV = nRT$; $\qquad n_{\text{final}} = n_{\text{initial}} + 1$ mole

$n_{\text{initial}} = \dfrac{PV}{RT} = \dfrac{(1\text{ atm})(1.35\text{ L})}{(273\text{ K})(0.0821\ \dfrac{\text{L} \cdot \text{atm}}{\text{K} \cdot \text{mole}})} = 0.06$ mole

$n_{final} = 0.16$ mole

$$V = \frac{nRT}{P} = \frac{(0.16 \text{ mole})(0.0821 \frac{L \cdot atm}{K \cdot mole})(273 \text{ K})}{1 \text{ atm}} = 3.59 \text{ L}$$

5.27 $8 \text{ g } O_2 \times \frac{1 \text{ mole } O_2}{32 \text{ g } O_2} = 0.25 \text{ mole } O_2$

$$V = \frac{nRT}{P} = \frac{(0.25 \text{ mole})(0.0821 \frac{L \cdot atm}{K \cdot mole})(273 \text{ K})}{1 \text{ atm}} = 5.60 \text{ L}$$

5.29 $P = 1$ atm.; $T = 273$ K; $R = 0.0821 \frac{L \cdot atm}{mole \cdot K}$

$V(NH_3, \text{ ammonia}) = 5.60$ L; $V(H_2, \text{ hydrogen}) = 11.2$ L

Ammonia → $n = \frac{PV}{RT} = \frac{(1 \text{ atm})(5.60 \text{ L})}{(273 \text{ K})(0.0821 \frac{L \cdot atm}{K \cdot mole})} = 0.25 \text{ mole} \times \frac{17 \text{ g } NH_3}{1 \text{ mole } NH_3} = 4.25 \text{ g}$

Hydrogen → $n = \frac{PV}{RT} = \frac{(1 \text{ atm})(11.2 \text{ L})}{(273 \text{ K})(0.0821 \frac{L \cdot atm}{K \cdot mole})} = 0.5 \text{ mole} \times \frac{2 \text{ g } H_2}{1 \text{ mole } H_2} = 1.00 \text{ g}$

5.31 $P_{total} = P_{water} + P_{oxygen} = 723$ torr

$P_{water} = 23.8$ torr

$P_{oxygen} = 723 - 23.8 = 699.2$ torr

$699.2 \text{ torr} \times \frac{1 \text{ atm}}{760 \text{ torr}} = 0.92 \text{ atm}$

5.33 $PV = nRT$; Watch your units here!

$$n = \frac{PV}{RT} = \frac{(752 \text{ torr})(\frac{1 \text{ atm}}{760 \text{ torr}})(6.00 \text{ L})}{(298 \text{ K})(0.0821 \frac{L \cdot atm}{K \cdot mole})} = 0.243 \text{ mole}$$

5.35

$$n_{water} = 18.0 \text{ g} \times \frac{1 \text{ mole H}_2\text{O}}{18.0 \text{ g H}_2\text{O}} = 1 \text{ mole} \qquad n_{H_2} = 2.02 \text{ g} \times \frac{1 \text{ mole H}_2}{2.02 \text{ g H}_2} = 1 \text{ mole}$$

$$n_{CO_2} = 44.0 \text{ g} \times \frac{1 \text{ mole CO}_2}{44.0 \text{ g CO}_2} = 1 \text{ mole}$$

Temperature and pressure are the same for the three containers; therefore:

$$V_{H_2O} = V_{H_2} = V_{CO_2} = \frac{nRT}{P} = \frac{(1 \text{ mole})(0.0821 \frac{L \cdot atm}{K \cdot mole})(273 \text{ K})}{1 \text{ atm}} = 22.4 \text{ L}$$

5.37 $P_{total} = P_{O_2} + P_{N_2} = 800 \text{ torr}$

$$P_{O_2} = x_{O_2} P_{total} \text{ where } x_{O_2} = \frac{n_{O_2}}{n_{O_2} + n_{N_2}}$$

$n_{O_2} = 0.250; n_{N_2} = 0.750$

Therefore, $P_{O_2} = 0.250(800 \text{ torr}) = 200 \text{ torr}$.

5.39 $$V = \frac{nRT}{P} = \frac{(1.50 \text{ mole})(0.0821 \frac{L \cdot atm}{K \cdot mole})(328 \text{ K})}{1 \text{ atm}} = 40.4 \text{ L}$$

5.41 $$\frac{P_1 V_1}{T_1} = \frac{P_2 V_2}{T_2} \rightarrow \frac{(565 \text{ torr})(800 \text{ mL})}{298 \text{ K}} = \frac{(760 \text{ torr})(850 \text{ mL})}{T_2}$$

As long as the dimensions of the initial and final volumes and pressures are the same, this equation can be used directly.

$T_2 = 425 \text{ K} = 153 \text{ °C}$

5.43 $$\frac{P_1 V_1}{T_1} = \frac{P_2 V_2}{T_2} \rightarrow \frac{(600 \text{ torr})(2.00 \text{ L})}{298 \text{ K}} = \frac{(783 \text{ torr}) V_2}{398 \text{ K}}$$

$V_2 = 2.0 \text{ L}$

5.45 $\dfrac{P_1V_1}{T_1} = \dfrac{P_2V_2}{T_2} \to \dfrac{(550 \text{ torr})V_1}{323 \text{ K}} = \dfrac{(850 \text{ torr})(1550 \text{ mL})}{860 \text{ K}}$

$V_1 = 900 \text{ mL}$

5.47 Density $= \dfrac{\text{mass}}{\text{volume}}$; mass = number of moles × molecular weight $\left[\dfrac{g}{\text{mole}}\right]$

Density $= \dfrac{n \times \text{molecular weight}}{\text{volume}} = \dfrac{n \times MW}{V} = \dfrac{P \times MW}{RT}$

Density $= \dfrac{(850 \text{ torr} \times \dfrac{1 \text{ atm}}{760 \text{ torr}})}{(303 \text{ K})(0.0821 \dfrac{L \cdot atm}{K \cdot \text{mole}})} \times \dfrac{2.02 \text{ g}}{\text{mole}} = 9.06 \times 10^{-2} \text{ g/L}$

5.49 $Zn + 2 \text{ HCl} \to ZnCl_2 + H_2$

13.1 g Zn consumed $\times \dfrac{1 \text{ mole}}{65.4 \text{ g}} = 0.2 \text{ mole Zn} \times \dfrac{1 \text{ mole } H_2}{1 \text{ mole Zn}} = 0.2 \text{ mole } H_2$

$V = \dfrac{nRT}{P} = \dfrac{(0.2 \text{ mole})(0.0821 \dfrac{L \cdot atm}{K \cdot \text{mole}})(273 \text{ K})}{1 \text{ atm}} = 4.49 \text{ L}$

$4.49 \text{ L} \times \dfrac{1000 \text{ mL}}{L} = 4490 \text{ mL}$

5.51 (a) $0.75 \text{ atm} \times \dfrac{760 \text{ torr}}{1 \text{ atm}} = 570 \text{ torr}$

(b) $1.23 \text{ atm} \times \dfrac{760 \text{ torr}}{1 \text{ atm}} = 935 \text{ torr}$

(c) $0.950 \text{ atm} \times \dfrac{760 \text{ torr}}{1 \text{ atm}} = 722 \text{ torr}$

(d) $1.750 \text{ atm} \times \dfrac{760 \text{ torr}}{1 \text{ atm}} = 1330 \text{ torr}$

5.53 Use $PV = nRT$ to get:

(a) 27.7 L (b) 23.1 atm (c) 329 K (d) 1.00 mole

5.55 $\dfrac{P_1}{n_1} = \dfrac{P_2}{n_2}$

$n_1 = \dfrac{P_1 V_1}{RT_1} = \dfrac{(14.0 \text{ atm})(20.0 \text{ L})}{(300 \text{ K})(0.0821 \dfrac{\text{L} \cdot \text{atm}}{\text{K} \cdot \text{mole}})} = 11.37 \text{ mole}$

Because T and V remain the same, $n_2 = \left(\dfrac{11.5 \text{ atm}}{14.0 \text{ atm}}\right)(11.37 \text{ mol}) = 9.34 \text{ mol}$

Therefore, the amount of gas that was removed from the cylinder is
11.37 mol − 9.34 mol = 2.03 mol

5.57 $C_H = \dfrac{0.51 \text{ mL } CO_2}{\text{mL blood}}$ at 1 atm

$\dfrac{V_{CO_2}}{V_{blood}} = C_H P_{CO_2} \rightarrow V_{CO_2} = \left(\dfrac{0.51 \text{ mL } CO_2}{\text{mL blood}}\right)(1 \text{ mL blood})\left(\dfrac{46 \text{ torr}}{760 \text{ torr}}\right) = 0.031 \text{ mL } CO_2$

Chapter 6
Interactions Between Molecules

6.1   All matter would exist only in the gaseous state. No matter would be present in either liquid or solid form.

6.3   The melting of a solid describes the transition from a structure consisting of a highly ordered array of molecules to a disordered structure with the molecules still in contact but now free to move about. This transition is effected by adding energy to the system to overcome the strong attractive forces in the solid.

6.5   True. The vapor pressure of a liquid is a measure of the escaping tendency of its molecules. Increasing the temperature increases the kinetic energy of its molecules, thus increasing their escaping tendency.

6.7   Because the molecules are in contact, there is no possibility of a change in volume with an increase in pressure.

6.9   The molar heat of vaporization is always greater than the molar heat of fusion because to melt a solid, only a part of the attractive forces must be overcome; whereas to vaporize a liquid, all secondary attractive forces must be overcome.

6.11 True. The transformation between the solid state and the liquid state occurs at the same temperature. If the transformation is approached by raising the temperature of the solid until the liquid forms, the temperature is called the melting point. If the transformation is approached by lowering the temperature of the liquid until the solid forms, the temperature is called the freezing point.

6.13 9.7 kcal of heat will be evolved because the condensation of a gas is merely the reverse of vaporization.

6.15 Through the weak London forces (temporary dipoles).

6.17 Hydrogen bonds require an H covalently bonded to an O, N, or F. Part b is the only pair in which both compounds meet this requirement.

6.19 The greater the number of electrons (atomic number) in a molecule, the larger the temporary dipole (London force). This means that, in general, the greater the molecular mass, the greater the strength of London forces exerted between molecules.

6.21 The four C-Cl polar bonds are arranged symmetrically (tetrahedrally) with carbon at the center of the tetrahedron. The symmetric arrangement of dipoles results in a net dipole moment of zero.

Interactions Between Molecules 33

6.23

	London force	Dipole–dipole	Hydrogen bond
CH_4	Yes	No	No
$CHCl_3$	Yes	Yes	No
NH_3	Yes	Yes	Yes

6.25 (a) has the greater heat of vaporization because it can form hydrogen bonds between molecules, whereas (b) cannot.

6.27 The molecules of a liquid have acquired sufficient kinetic energy to partly overcome the intermolecular forces holding them in fixed positions. Although they are free to assume new positions under an applied force (for example, gravity), they still remain in contact.

6.29 In an open system, vapor molecules will continuously escape into the surroundings. Therefore the rate of evaporation will always exceed the rate of condensation, and evaporation will occur continuously.

6.31 When a solid melts, the ordered solid-state structure becomes disordered, but the molecules in the liquid state remain in close contact. Therefore the volume of the liquid is close to that of the solid.

6.33 The word "vapor" is used to describe the gaseous state of a substance whose liquid and gaseous states are present at the same time. Under normal conditions of room temperature, the liquid states of oxygen or nitrogen are not possible.

6.35 c < a < b. These compounds are ordered by increasing strength of intermolecular forces: dipole–dipole forces < hydrogen bond < ionic.

6.37 No. Vapor pressure is a balance between evaporation and condensation. In an open container, the rate of evaporation is always greater than the rate of condensation, and equilibrium cannot be established.

6.39 Equilibrium describes a situation that does not change over time.

6.41 The temperature at which vapor bubbles form throughout a liquid at an external pressure of 1 atmosphere.

6.43 Both molecules are polar, but, additionally, water can form hydrogen bonds, so the secondary attractive forces in water are greater than those in chloroform. Therefore water will have the greater surface tension of the two.

6.45 Acetic acid molecules form hydrogen bonds among themselves and to water. Pentane molecules can interact only by London forces. The attractive forces between acetic acid and water are similar, and therefore acetic acid will dissolve in water. The attractive forces

34 Chapter 6

between pentane molecules and water are very different and will therefore not form solutions.

6.47 An amorphous solid has no particular shape, melts over a range of temperatures, and, if it can be shattered, forms pieces that have round or smooth surfaces resembling a liquid

6.49 $CH_4 \rightarrow C_2H_5OH \rightarrow NaCl$. These are ordered by decreasing strength of intermolecular forces: London dispersion < hydrogen bond < ionic forces.

6.51 $\dfrac{21.6}{31.8} = 0.68 = 68\%$

6.53 A molecule that has polar bonds and is not symmetric will exhibit a dipole. Examples include H_2O, HF, and $CHCl_3$.

6.55 H_2O can form two hydrogen bonds per molecule, whereas HF can form only one hydrogen bond per molecules. Therefore, more thermal energy is required to break two hydrogen bonds than one. This results in a higher boiling point.

6.57 Vapor pressure measures the tendency of molecules to escape from the liquid into the gas phase. The greater the secondary forces, the lower the escaping tendency and the smaller the vapor pressure.

6.59 The Lewis dot structure of dimethyl ether shows that the oxygen has a free pair of electrons. This free pair of electrons acts to force the molecules into the bent configuration like that of water. Thus, dimethyl ether has a dipole.

Chapter 7

Solutions

7.1 (a) $25.2 \text{ g solution} \times \dfrac{4.25 \text{ g solute}}{100 \text{ g solution}} = 1.07 \text{ g solute}$

(b) $125 \text{ g solution} \times \dfrac{6.055 \text{ g solute}}{100 \text{ g solution}} = 8.19 \text{ g solute}$

7.3 0.600% w/v solution means $\dfrac{\text{grams of solute}}{\text{mL of solution}} = \dfrac{0.600}{100} = 0.006$

$\dfrac{x \text{ g KHCO}_3}{250 \text{ mL solution}} = 0.006 \rightarrow x = 1.50 \text{ g KHCO}_3$

7.5 6.20% w/v solution means $\dfrac{\text{grams of solute}}{\text{mL of solution}} = \dfrac{6.20}{100} = 0.062$

$\dfrac{5.4 \text{ g NH}_4\text{Cl}}{x \text{ mL solution}} = 0.062 \rightarrow x = 87.1 \text{ mL of solution}$

7.7 5.00% w/w solution means $\dfrac{\text{grams of solute}}{\text{grams of solution}} = \dfrac{5.00}{100} = 0.05$

$\dfrac{6.25 \text{ g glucose}}{x \text{ g solution}} = 0.05 \rightarrow x = 125 \text{ g of solution}$

7.9 7.50% v/v solution means $\dfrac{\text{volume of solute}}{\text{volume of solution}} = \dfrac{7.50}{100} = 0.075$

$\dfrac{33.6 \text{ mL propylene glycol}}{x \text{ mL solution}} = 0.075 \rightarrow x = 448 \text{ mL of solution}$

7.11 mg % solution means $\dfrac{\text{mg of solute}}{\text{mL of solution}} \times 100$

$\dfrac{0.230 \text{ mg}}{19.0 \text{ mL}} \times 100 = 1.21 \text{ mg \%}$

36   Chapter 7

7.13 $42 \text{ ppm} = \dfrac{42 \text{ mg of solute}}{1 \text{ L of solution}}$ $\% \text{ w/v} = \dfrac{\text{g of solute}}{\text{mL of solution}} \times 100$

$42 \text{ mg} \times \dfrac{1 \text{ g}}{1000 \text{ mg}} = 0.042 \text{ g}$ $1 \text{ L} \times \dfrac{1000 \text{ mL}}{1 \text{ L}} = 1000 \text{ mL}$

$\dfrac{0.042 \text{ g}}{1000 \text{ mL}} \times 100 = 0.0042\% \text{ w/v}$

7.15

2.5 L of a 0.40 M solution = $2.5 \text{ L} \times \dfrac{0.40 \text{ mol of solute}}{1 \text{ L of solution}} = 1.0$ mol of solute

7.17   Molarity = [M] = $\dfrac{\text{moles of solute}}{\text{L of solution}}$

(a) $\dfrac{2.6 \text{ mol of solute}}{1.3 \text{ L of solution}} = 2.0 \text{ M}$

(b) $\dfrac{0.810 \text{ mol of solute}}{0.240 \text{ L of solution}} = 3.38 \text{ M}$

7.19   Moles of solute = (L of solution) × (molarity)

(a) (1.05 L) × (2.65 M) = 2.78 mol

(b) (0.452 L) × (0.850 M) = 0.384 mol

7.21   Molarity = [M] = $\dfrac{\text{moles of solute}}{\text{L of solution}}$; therefore volume of solution = $\dfrac{\text{moles of solute}}{\text{molarity}}$

(a) $\dfrac{1.35 \text{ mol of solute}}{0.900 \text{ M}} = 1.50$ L of solution

(b) $\dfrac{2.8 \text{ mol of solute}}{0.731 \text{ M}} = 3.83$ L of solution

7.23 $23.83 \text{ mg MgCl}_2 \times \dfrac{1 \text{ g}}{1000 \text{ mg}} \times \dfrac{1 \text{ mol}}{95.3 \text{ g}} = 0.00025 \text{ mol MgCl}_2$

$\dfrac{0.00025 \text{ mol of solute}}{0.275 \text{ L of solution}} = 9.09 \times 10^{-4} \text{ M}$

7.25 $0.450 \text{ M} \times 0.075 \text{ L} = 0.0337 \text{ mol CaCl}_2 \times \dfrac{111.1 \text{ g}}{\text{mole}} \times \dfrac{1000 \text{ mg}}{\text{g}} = 3.75 \times 10^3 \text{ mg}$

7.27 $25.0 \text{ g KCl} \times \dfrac{1 \text{ mol}}{74.6 \text{ g}} = 0.33 \text{ mol}$

$\dfrac{0.33 \text{ mol KCl}}{0.620 \dfrac{\text{mol KCl}}{\text{L}}} = 0.54 \text{ L} \times \dfrac{1000 \text{ mL}}{\text{L}} = 540 \text{ mL}$

7.29 $V_1 M_1 = V_2 M_2$

$V_1 = ?$ $M_1 = 36 \text{ M}$

$V_2 = 6.0 \text{ L}$ $M_2 = 0.12 \text{ M}$

$V_1 = \dfrac{6.0 \text{ L} \times 0.12 \text{ M}}{36 \text{ M}} = 0.020 \text{ L} = 20 \text{ mL}$

7.31 $V_1 M_1 = V_2 M_2$

(a) $V_1 = ?$ $M_1 = 0.750 \text{ M}$

$V_2 = 4.50 \text{ L}$ $M_2 = 0.250 \text{ M}$

$V_1 = \dfrac{4.50 \text{ L} \times 0.250 \text{ M}}{0.750 \text{ M}} = 1.50 \text{ L}$

(b) $V_1 = ?$ $M_1 = 0.360 \text{ M}$

$V_2 = 650 \text{ mL}$ $M_2 = 0.18 \text{ M}$

$V_1 = \dfrac{650 \text{ mL} \times 0.18 \text{ M}}{0.36 \text{ M}} = 325 \text{ mL}$

7.33 $V_1 M_1 = V_2 M_2$

$V_1 = 0.80 \text{ L}$ $M_1 = ?$

$V_2 = 4.0 \text{ L}$ $M_2 = 1.8\% \text{ w/v}$

$M_1 = \dfrac{4.0 \text{ L} \times 1.8\% \text{ w/v}}{0.80 \text{ L}} = 9.0\% \text{ w/v}$

38   Chapter 7

7.35 $V_1M_1 = V_2M_2$

$V_1 = 0.1$ L $M_1 = 2.10$ M

$V_2 = 0.420$ L $M_2 = ?$

$M_2 = \dfrac{0.1 \text{ L} \times 2.10 \text{ M}}{0.420 \text{ L}} = 0.500$ M

7.37   No. If the solution at 10°C were saturated, it would be at a concentration on 43 g/100 mL. When the temperature of that solution is raised to 50°C, the largest amount of strontium acetate that could stay in solution is 37.4 g/100 mL. Therefore, to attain this concentration, 5.6 g/100 mL of strontium acetate would have to precipitate out of solution. Because the solution remained clear, it must not have been saturated at 10°C.

7.39   Molarity = [M] = $\dfrac{\text{moles of solute}}{\text{L of solution}}$

$74.5 \text{ g} \times \dfrac{1 \text{ mol}}{111.1 \text{ g}} = 0.671$ mol $100 \text{ mL} \times \dfrac{1 \text{ L}}{1000 \text{ mL}} = 0.1$ L

$\dfrac{0.671 \text{ mol}}{0.1 \text{ L}} = 6.71$ M

7.41   (a) $AlCl_3$ ionizes to $Al^{3+} + 3\ Cl^-$; therefore i = 4; 0.1 M × 4 = 0.4 osmol/L

       (b) KCl ionizes to $K^+ + Cl^-$; therefore i = 2; 0.01 M × 2 = 0.02 osmol/L

7.43   As stated in Box 7.3 a solution must be 0.3 osmol/L to be isotonic with red blood cells.

8.0% w/v glucose solution = $\dfrac{8 \text{ g}}{100 \text{ mL}}$ NaCl solution

Glucose does not dissociate in solution.

$\dfrac{8 \text{ g glucose}}{100 \text{ mL}} \times \dfrac{1000 \text{ mL}}{\text{L}} \times \dfrac{1 \text{ mol}}{180 \text{ g}} = 0.44 \dfrac{\text{mol}}{\text{L}} \times \dfrac{1 \text{ ion}}{\text{mol}} = 0.44 \dfrac{\text{osmol}}{\text{L}}$

0.44 osmol/L > 0.3 osmol/L. Therefore water will flow from the cell into the solution, the erythrocytes will lose water and can shrink.

7.45   (a) $\dfrac{13.0 \text{ g CaCl}_2}{13.0 \text{ g} + 133 \text{ g H}_2\text{O}} = 0.089 \times 100 = 8.90\%$ w/w.

(b) $\dfrac{12.5 \text{ g ethyl alcohol}}{12.5 \text{ g} + 165 \text{ g H}_2\text{O}} = 0.0704 \times 100 = 7.04\%$ w/w

(c) $\dfrac{2.5 \text{ g}}{2.5 \text{ g} + 60 \text{ g}} = 0.04 \times 100 = 4.0\%$ w/w

7.47 (a) $55.0 \text{ g of solution} \times \dfrac{4 \text{ g glucose}}{100 \text{ g of solution}} = 2.2 \text{ g of glucose}$

(b) $175 \text{ g of solution} \times \dfrac{7.5 \text{ g ammonium sulfate}}{100 \text{ g of solution}} = 13.1 \text{ g of ammonium sulfate}$

7.49 (a) $\dfrac{6.4 \text{ g}}{66.0 \text{ mL}} \times 100 = 9.70\%$ w/v (b) $\dfrac{1.3 \text{ g}}{125 \text{ mL}} \times 100 = 1.04\%$ w/v

(c) $\dfrac{3.6 \text{ g}}{90.0 \text{ mL}} \times 100 = 4.00\%$ w/v

7.51 Molarity = [M] = $\dfrac{\text{moles of solute}}{\text{L of solution}}$

Therefore number of moles in solution = (molarity) × (L of solution).

(a) 0.066 M × 0.046 L = 0.003 moles (b) 0.065 M × 1.2 L = 0.078 moles

(c) 0.25 M × 0.495 L = 0.124 moles (d) 1.75 M × 0.625 L = 1.09 moles

7.53 0.375 M = $\dfrac{0.375 \text{ moles of glucose}}{\text{L of solution}}$

(a) $\dfrac{0.375 \text{ moles of glucose}}{\text{L of solution}} \times 1.49 \text{ L} = 0.559 \text{ moles}$

(b) $\dfrac{0.375 \text{ moles of glucose}}{\text{L of solution}} \times 6.72 \text{ L} = 2.52 \text{ moles}$

(c) $\dfrac{0.375 \text{ moles of glucose}}{\text{L of solution}} \times 2.56 \text{ L} = 0.960 \text{ moles}$

(d) $\dfrac{0.375 \text{ moles of glucose}}{\text{L of solution}} \times 1.81 \text{ L} = 0.679 \text{ moles}$

40   Chapter 7

7.55 $0.160 \text{ M} = \dfrac{0.160 \text{ moles of KI}}{\text{L of solution}} \times 0.5 \text{ L} = 0.08 \text{ moles KI needed}$

$0.08 \text{ moles KI} \times \dfrac{166 \text{ g}}{1 \text{ mole}} = 13.3 \text{ g of KI needed}$

Therefore, add water to 13.3 g of KI to a final volume of 500.0 mL.

7.57 (a) $\dfrac{15.0 \text{ g CaCl}_2}{15.0 \text{ g} + 250 \text{ g H}_2\text{O}} = 0.0566 \times 100 = 5.66\% \text{ w/w}$

(b) $\dfrac{16.5 \text{ g NaCl}}{16.5 \text{ g} + 325 \text{ g H}_2\text{O}} = 0.0483 \times 100 = 4.83\% \text{ w/w}$

(c) $\dfrac{2.20 \text{ g C}_2\text{H}_5\text{OH}}{2.20 \text{ g} + 122 \text{ g H}_2\text{O}} = 0.0177 \times 100 = 1.77\% \text{ w/w}$

7.59 (a) $0.950 \text{ L} \times \dfrac{1.25 \text{ mol NaCl}}{\text{L}} \times \dfrac{58.49 \text{ g}}{1 \text{ mol}} = 69.5 \text{ g NaCl}$

(b) $625 \text{ mL} \times \dfrac{1 \text{ L}}{1000 \text{ mL}} \times \dfrac{0.750 \text{ mol Mg(NO}_3)_2}{\text{L}} \times \dfrac{148.2 \text{ g}}{1 \text{ mol}} = 69.5 \text{ g Mg(NO}_3)_2$

(c) $325 \text{ mL} \times \dfrac{1 \text{ L}}{1000 \text{ mL}} \times \dfrac{2.20 \text{ mol K}_2\text{CO}_3}{\text{L}} \times \dfrac{138.2 \text{ g}}{1 \text{ mol}} = 98.7 \text{ g K}_2\text{CO}_3$

7.61 $V_1 M_1 = V_2 M_2$

$V_1 = 1.50 \text{ L} \qquad\qquad M_1 = 1.80 \text{ M}$

$V_2 = ? \qquad\qquad M_2 = 0.180$

$V_2 = \dfrac{1.50 \text{ L} \times 1.80 \text{ M}}{0.180 \text{ M}} = 15.0 \text{ L}$

7.63 $V_1 M_1 = V_2 M_2$

$V_1 = 0.150 \text{ L} \qquad\qquad M_1 = ?$

$V_2 = 1.50 \text{ L} \qquad\qquad M_2 = 0.90\% \text{ w/v}$

$M_1 = \dfrac{1.5 \text{ L} \times 0.9\% \text{ w/v}}{0.150 \text{ L}} = 9.0\% \text{ w/v}$

7.65 From Table 7.3:

(a) AgCl is not soluble in H₂O; therefore a precipitate forms:

$$Ag^+ (aq) + Cl^- (aq) \rightarrow AgCl (s)$$

(b) Mg(OH)₂ is not soluble in H₂O; therefore a precipitate forms:

$$Mg^{2+} (aq) + 2\ OH^- (aq) \rightarrow Mg(OH)_2 (s)$$

(c) PbCl₂ is not soluble in H₂O; therefore a precipitate forms:

$$Pb^{2+} (aq) + 2\ Cl^- (aq) \rightarrow PbCl_2 (s)$$

(d) Ca₃(PO₄)₂ is not soluble in H₂O; therefore a precipitate forms:

$$3\ Ca^{2+} (aq) + 2\ PO_4^{3-} (aq) \rightarrow Ca_3(PO_4)_2 (s)$$

7.67 In a solution, the minor component is the solute and the major component is the solvent. Therefore, propylene glycol is the solvent and water is the solute.

7.69 100 torr > 15 torr; therefore water moves out of the arterial blood.

7.71 From box 7.3, 0.30 osmol/L.

Chapter 8
Chemical Reactions

8.1 $3 O_2 (g) \rightarrow 2 O_3 (g)$

The ratio of O_3 to O_2 is $\dfrac{2 \text{ mol } O_3}{3 \text{ mol } O_2}$

$$\dfrac{1.50 \text{ mol } O_3}{1 \text{ min}} \times \dfrac{3 \text{ mol } O_2}{2 \text{ mol } O_3} = \dfrac{2.25 \text{ mol } O_2 \text{ (g)}}{\text{min}}$$

8.3   The increase in temperature results in a proportional increase in the number of molecules that undergo collisions that have enough energy to result in reaction.

8.5   For reaction a to take place, the two reactants must have kinetic energies (K.E.) greater than 24 kJ; whereas for reaction b to take place, the two reactants must have kinetic energies greater than 53 kJ. At a given temperature, the percentage of molecules that have K.E. > 24 kJ will be greater than the percentage that have K.E. > 53 kJ. Therefore, the rate of reaction a will be greater than the rate of reaction b.

8.7 $\Delta H_{reaction} = E_{forward} - E_{back}$

$-21 \text{ kJ} = 37 \text{ kJ} - E_{back}$

$E_{back} = 58 \text{ kJ}$

8.9   No. The reaction takes place in an open system, so the CO_2 escapes to the atmosphere. Therefore, the CO_2 that is formed cannot react with CaO to allow the reverse reaction to take place. Therefore, the system can never come to equilibrium.

8.11   Forward reaction: $CaCO_3 (s) \rightarrow CaO (s) + CO_2 (g)$

Reverse reaction: $CaO (s) + CO_2 (g) \rightarrow CaCO_3 (s)$

8.13   Pure solids and liquids do not appear in equilibrium constants; therefore, $K_{eq} = [CO_2]$.

8.15 $K_{eq} = \dfrac{[HI]^2}{[H_2][I_2]} = \dfrac{(0.27)^2}{(0.86)(0.86)} = 0.099$

8.17   Le Chatelier's principle states that, when a reactant or product is removed from a reaction at equilibrium, the equilibrium shifts to replace the compound that was removed.

$$N_2(g) + 3 H_2(g) \leftrightarrow 2 NH_3$$

If NH_3 is removed, the reaction shifts to replace NH_3, which means that it shifts to the products.

8.19 (a) Le Chatelier's principle states that, when the temperature is lowered in a reaction at equilibrium, the equilibrium shifts to raise the temperature. In this reaction, that means that the reaction shifts to the reactants, which will result in a color shift to pink.

(b) Le Chatelier's principle states that, when a reactant or product is added to a reaction at equilibrium, the equilibrium shifts to use up the compound that was added. In this reaction, when Cl^- is added, the reaction shifts to use up more Cl^-; the reaction will shift to the products, which will result in a color shift to blue.

8.21 The square brackets around the chemical compounds in an equilibrium constant expression denote the molar concentration at equilibrium.

8.23 The concentration of pure liquids and solids do not appear in K_{eq}; therefore:

(a) $K_{eq} = \dfrac{[CO][H_2]^3}{[CH_4][H_2O]}$ (b) $K_{eq} = \dfrac{[CO_2]^8[H_2O]^8}{[O_2]^{12}}$

8.25 The concentration of pure liquids and solids do not appear in K_{eq}; therefore:

(a) $K_{eq} = [CO_2][NH_3]^2$ (b) $K_{eq} = \dfrac{1}{[O_2]^{0.5}}$

8.27 (a) $COCl_2(g) \leftrightarrow CO(g) + Cl_2(g)$

(b) $CH_4(g) + H_2O(g) \leftrightarrow CO(g) + 3 H_2(g)$

8.29 The concentration of pure liquids and solids do not appear in K_{eq}; therefore:

(a) $K_{eq} = \dfrac{[NO_2][NO_3]}{[N_2O_5]}$ (b) $K_{eq} = \dfrac{[N_2]^2[H_2O]^6}{[NH_3]^4[O_2]^3}$

8.31 $A(g) \leftrightarrow B(g)$

$K_{eq} = \dfrac{[B]}{[A]}$; therefore at equilibrium $\dfrac{[A]}{[B]} = \dfrac{1}{K_{eq}}$

(a) 1000/1 (b) 1/100 (c) 7.1/1 (d) 12/1

8.33 The dissolution of NH_4CNS can be thought of as the reaction

$$NH_4CNS\ (s) \leftrightarrow NH_4CNS\ (aq)$$

The solution becomes cold when this "reaction" goes to the right; therefore heat is needed to allow this reaction to take place.

$$NH_4CNS\ (s) + heat \leftrightarrow NH_4CNS\ (aq)$$

If the temperature is raised, Le Chatelier's principle tells us that the reaction will shift to lower the temperature. In this reaction, that means that the reaction shifts to the right and more solid will dissolve.

8.35 $\Delta H_{reaction} = E_{forward} - E_{back}$

15 kJ = $E_{forward}$ − 57 kJ

$E_{forward}$ = 72 kJ

8.37 $\Delta H_{reaction} = E_{forward} - E_{back}$

Because $\Delta H > 0$, $E_{forward} > E_{back}$

8.39 If a system in an equilibrium state is disturbed, the system will adjust to neutralize that disturbance and restore the system to equilibrium.

Chapter 9

Acids, Bases, and Buffers

9.1 The ion product: $K_w = [H_3O^+] \times [OH^-] = (1.00 \times 10^{-7}) \times (1.00 \times 10^{-7}) = 1.00 \times 10^{-14}$

9.3 An amphoteric substance is a compound that can act as an acid or a base.

9.5 A strong base is ionized 100% in aqueous solution. Two examples are NaOH and LiOH.

9.7 0.30 M HCl solution. HCl is a strong acid, which means that every HCl molecule dissociates into an H_3O^+ ion and a Cl^- ion. Thus, $[H_3O^+] = [Cl^-] = 0.30$ M

9.9 0.25 M TlOH solution. TlOH is a strong base, which means that every TlOH molecule dissociates into an Tl^+ ion and an OH^- ion. Thus, $[Tl^+] = [OH^-] = 0.25$ M

9.11 0.350 M HCl solution. HCl is a strong acid, which means that every HCl molecule dissociates into an H_3O^+ ion and a Cl^- ion. Thus, $[H_3O^+] = [Cl^-] = 0.350$ M. It is also known that

$$[H_3O^+] \times [OH^-] = 1 \times 10^{-14} \rightarrow [OH^-] = \frac{1 \times 10^{-14}}{0.35} = 2.86 \times 10^{-14}$$

9.13 0.020 M Ca(OH)$_2$ solution. Ca(OH)$_2$ is a strong base, which means that every Ca(OH)$_2$ molecule dissociates into an Ca^{2+} ion and two OH^- ions. Thus, $[Ca^{2+}] = 0.020$ M. Because there are twice as many OH^- ions in solution as Ca^{2+} ions, their concentration must be twice that of Ca^{2+}. Thus, $[OH^-] = 0.020$ M $\times 2 = 0.040$ M. It is also known that

$$[H_3O^+] \times [OH^-] = 1 \times 10^{-14} \rightarrow [H_3O^+] = \frac{1 \times 10^{-14}}{0.04} = 2.5 \times 10^{-13}$$

9.15 The pH of pure water at 25°C is 7.00.

9.17 pH = –log $[H_3O^+]$ = – log (0.01) = 2.0

9.19 pOH = – log$[OH^-]$ = –log(0.015) = 1.82

$$[H_3O^+] \times [OH^-] = 1 \times 10^{-14} \rightarrow [H_3O^+] = \frac{1 \times 10^{-14}}{0.015} = 6.67 \times 10^{-13}$$

pH = –log $[H_3O^+]$ = – log (6.67 × 10^{-13}) = 12.18

Alternatively, pH + pOH = 14; thus pH = 14 – 1.82 = 12.18.

46   Chapter 9

9.21   Each of the acids in this question is a strong acid, which means that every acid molecule dissociates into an H_3O^+ ion and a negative counter ion. Thus, $[H_3O^+]$ will equal the concentration of the acid. Thus

(a) $pH = -\log[H_3O^+] = -\log(0.00336) = 2.46$

(b) $pH = -\log[H_3O^+] = -\log(0.0253) = 1.60$

(c) $pH = -\log[H_3O^+] = -\log(0.000730) = 3.14$

(d) $pH = -\log[H_3O^+] = -\log(0.0740) = 1.13$

9.23   A weak acid is one that incompletely dissociates in water. Examples are given in Table 9.4; acetic acid and carbonic acid.

9.25   Because the pK_a value of the weak acid, acetic acid, and the pK_b value of the weak base, ammonia, are approximately equal, both will be dissociated in aqueous solution to a similar extent. Therefore, the OH^- ions produced by the reaction of ammonia with water will neutralize all of the $[H_3O^+]$ ions created by acetic acid and vice versa; thus the solution will be neutral.

9.27   (a) An acid is a compound that donates a proton; its conjugate base is the compound that is formed on loss of a hydrogen. Both HNO_2 and H_3O^+ give up a proton to form NO_2^- and H_2O, respectively. Therefore HNO_2 and NO_2^- are one conjugate acid-base pair, and H_3O^+ and H_2O are another.

(b) An acid is a compound that donates a proton; its conjugate base is the compound that is formed on loss of a hydrogen. Both $H_2PO_4^-$ and H_3O^+ give up a proton to form HPO_4^{2-} and H_2O, respectively. Therefore $H_2PO_4^-$ and HPO_4^{2-} are one conjugate acid-base pair, and H_3O^+ and H_2O are another.

9.29   The equilibrium constant for the reaction of acetate ion with water is $K_b = 5.75 \times 10^{-10}$.

9.31 $pK_a = -\log(K_a) = -\log(1.74 \times 10^{-5}) = 4.76$

$pK_b = -\log(K_b) = -\log(5.75 \times 10^{-10}) = 9.24$

9.33   From the solution to Exercise 9.31, $pK_a + pK_b = 4.76 + 9.24 = 14.00$

9.35 $H_2CO_3\ (aq) + H_2O\ (l) \leftrightarrow H_3O^+\ (aq) + HCO_3^-\ (aq)$

$HCO_3^-\ (aq) + H_2O\ (l) \leftrightarrow H_3O^+\ (aq) + CO_3^{2-}\ (aq)$

9.37 $K_{a1} = \dfrac{[H_3O^+][HCO_3^-]}{[H_2CO_3]} = 4.45 \times 10^{-7}$

Acids, Bases, and Buffers   47

$$K_{a2} = \frac{[H_3O^+][CO_3^{2-}]}{[HCO_3^-]} = 4.72 \times 10^{-11}$$

9.39 From the equations in the solution to Exercise 9.37, the amount of H_3O^+ contributed by HCO_3^- is a factor of 10,000 less than that contributed by H_2CO_3 ($10^{-11}/10^{-7} = 1/10,000$). Thus it will be of minor significance.

9.41 Salts composed of cations and anions of strong acids and bases form neutral solutions, whereas salts having anions of weak acids form basic solutions and salts having cations of weak bases form acidic solutions; thus

(a) $CaCl_2$ is a salt of a strong base and a strong acid, $Ca(OH)_2$ and HCl; thus the solution is neutral.

(b) Na_2CO_3 is the salt of the anion of a weak acid, H_2CO_3; thus the solution is basic.

(c) $FeCl_3$ is the salt of the cation of a weak base, $Fe(OH)_3$; thus the solution is acidic

(d) NH_4Cl is the salt of the cation of a weak base, NH_3; thus the solution is acidic.

(e) $MgSO_4$ is a salt of a strong base and a strong acid, $Mg(OH)_2$ and H_2SO_4; thus the solution is neutral.

9.43 Salts composed of cations and anions of strong acids and bases form neutral solutions, whereas salts having anions of weak acids form basic solutions and salts having cations of weak bases form acidic solutions; thus

(a) $Ca(NO_3)_2$ is a salt of a strong base and a strong acid, $Ca(OH)_2$ and HNO_3; thus the solution is neutral.

(b) $NaNO_2$ is the salt of the anion of a weak acid, HNO_2; thus the solution is basic.

(c) KCN is the salt of the anion of a weak acid, HCN; thus the solution is basic.

(d) CH_3COONa is the salt of the anion of a weak acid, CH_3COOH; thus the solution is basic.

(e) HCOOMg is the salt of the anion of a weak acid, HCOOH; thus the solution is basic.

9.45 The presence of the conjugate acid allows the solution to neutralize the addition of a base, whereas the presence of the conjugate base allows the solution to neutralize the addition of a base. Because it is a conjugate acid-base pair, the solution remains at a constant pH.

9.47 $pH = pK_a + \log\left(\dfrac{\text{proton acceptor}}{\text{proton donor}}\right) = \log\left(\dfrac{0.0080}{0.0060}\right) = 4.76 + 0.125 = 4.89$

48   Chapter 9

9.49 $\text{pH} = \text{p}K_a + \log\left(\dfrac{\text{proton acceptor}}{\text{proton donor}}\right) = \log\left(\dfrac{0.075}{0.050}\right) = 7.20 + 0.18 = 7.38$

9.51 $16.2 \text{ mL} \times \dfrac{1 \text{L}}{1000 \text{ mL}} \times \dfrac{0.021 \text{ mol}}{\text{L}} = 0.0003402 \text{ mol of KOH}$

0.0003402 mol KOH = 0.0003402 mol HCl when neutral

$\dfrac{0.0003402 \text{ mol}}{0.020 \text{L}} = 0.017 \text{M}$

9.53 $33.2 \text{ mL} \times \dfrac{1 \text{L}}{1000 \text{ mL}} \times \dfrac{0.041 \text{ mol}}{\text{L}} = 0.00136 \text{ mol KOH}$

$0.00136 \text{ mol KOH} \times \dfrac{1 \text{ mole } H_2SO_4}{2 \text{ mole KOH}} = 0.000680 \text{ mol } HNO_3 \text{ when neutral}$

$\dfrac{0.00068 \text{ moles}}{0.04 \text{L}} = 0.0170 \text{M}$

9.55   0.500 N solution denotes 0.5 equivalents of acid per liter. An equivalent is defined as the number of moles of dissociable H^+ per mole of acid. In H_3PO_4, three H^+ ions can come from a single H_3PO_4 molecule; thus

$\dfrac{0.5 \text{ equivalent}}{\text{L}} = \dfrac{x \text{ mole}}{\text{L}} \times \dfrac{3 \text{ moles } H^+}{\text{mole}}$ $x = 0.167$ moles H_3PO_4 per liter

0.167 moles $H_3PO_4 \times 98$ g/mole = 16.3 g H_3PO_4

Thus, 16.3 g of H_3PO_4 must be dissolved in sufficient water to make 1.00 L of solution.

9.57 $0.024 \text{ M } H_3PO_4 \times \dfrac{3 \text{ moles } H^+}{\text{mole}} = 0.072 \text{ N solution}$

9.59   (a) $0.125 \text{ N Na}^+ \times \dfrac{1 \text{ mole}}{1 \text{ eq of charge}} = 0.125 \text{ M}$

(b) $0.035 \text{ N K}^+ \times \dfrac{1 \text{ mole}}{1 \text{ eq of charge}} = 0.035 \text{ M}$

(c) $0.072 \text{ N Ca}^{2+} \times \dfrac{1 \text{ mole}}{2 \text{ eq of charge}} = 0.036 \text{ M}$

(d) $0.028 \text{ N Mg}^{2+} \times \dfrac{1 \text{ mole}}{2 \text{ eq of charge}} = 0.014 \text{ M}$

9.61 (a) 0.0031 M HNO$_3$ solution. HNO$_3$ is a strong acid, which means that every HNO$_3$ molecule dissociates into an H$_3$O$^+$ ion and a NO$_3^-$ ion. Thus, [H$_3$O$^+$] = 0.0031 M.

$$\text{pH} = -\log [\text{H}_3\text{O}^+] = -\log (0.0031) = 2.50$$

(b) 1.0 M HCl solution. HCl is a strong acid, which means that every HCl molecule dissociates into an H$_3$O$^+$ ion and a Cl$^-$ ion. Thus, [H$_3$O$^+$] = 1.0 M.

$$\text{pH} = -\log [\text{H}_3\text{O}^+] = -\log (1.0) = 0.0$$

(c) 0.0069 M HI solution. HI is a strong acid, which means that every HI molecule dissociates into an H$_3$O$^+$ ion and a I$^-$ ion. Thus, [H$_3$O$^+$] = 0.0069 M.

$$\text{pH} = -\log [\text{H}_3\text{O}^+] = -\log (0.0069) = 2.16$$

(d) 0.019 M HBr solution. HNO$_3$ is a strong acid, which means that every HBr molecule dissociates into an H$_3$O$^+$ ion and a Br$^-$ ion. Thus, [H$_3$O$^+$] = 0.0031 M.

$$\text{pH} = -\log [\text{H}_3\text{O}^+] = -\log (0.019) = 1.72$$

(e) 0.023 M HClO$_3$ solution. HClO$_3$ is a strong acid, which means that every HClO$_3$ molecule dissociates into an H$_3$O$^+$ ion and a ClO$_3^-$ ion. Thus, [H$_3$O$^+$] = 0.023 M.

$$\text{pH} = -\log [\text{H}_3\text{O}^+] = -\log (0.023) = 1.64$$

9.63 (a) 0.0062 M KOH solution. KOH is a strong acid, which means that every KOH molecule dissociates into an OH$^-$ ion and a K$^+$ ion. Thus, [OH$^-$] = 0.0062 M.

$$\text{pOH} = -\log [\text{OH}^-] = -\log (0.0062) = 2.21$$

(b) 0.0041 M Ca(OH)$_2$ solution. Ca(OH)$_2$ is a strong acid, which means that every Ca(OH)$_2$ molecule dissociates into two OH$^-$ ion and a Ca^{2+} ion. Thus, [OH$^-$] = 0.0082 M.

$$\text{pOH} = -\log [\text{OH}^-] = -\log (0.0082) = 2.09$$

(c) 0.028 M Ba(OH)$_2$ solution. Ba(OH)$_2$ is a strong acid, which means that every Ba(OH)$_2$ molecule dissociates into two OH$^-$ ion and a Ba^{2+} ion. Thus, [OH$^-$] = 0.056 M.

$$\text{pOH} = -\log [\text{OH}^-] = -\log (0.056) = 1.25$$

(d) 1.0 M KOH solution. KOH is a strong acid, which means that every KOH molecule dissociates into an OH$^-$ ion and a K$^+$ ion. Thus, [OH$^-$] = 1.0 M.

$$\text{pOH} = -\log[\text{OH}^-] = -\log(1.0) = 0.0$$

(e) 0.01 M NaOH solution. NaOH is a strong acid, which means that every NaOH molecule dissociates into an OH⁻ ion and a Na⁺ ion. Thus, [OH⁻] = 0.01 M.

$$\text{pOH} = -\log[\text{OH}^-] = -\log(0.01) = 2$$

9.65 Salts composed of cations and anions of strong acids and bases form neutral solutions, whereas salts having anions of weak acids form basic solutions and salts having cations of weak bases form acidic solutions; thus

(a) $Fe_2(SO_4)_3$ is the salt of the cation of a weak base, $Fe(OH)_3$; thus the solution is acidic.

(b) NaBr is a salt of a strong base and a strong acid, NaOH and HBr; thus the solution is neutral.

(c) $NaNO_2$ is the salt of the anion of a weak acid, HNO_2; thus the solution is basic.

(d) NH_4NO_3 is the salt of the cation of a weak base, NH_3; thus the solution is acidic.

(e) $Mg(CN)_2$ is the salt of the anion of a weak acid, HCN; thus the solution is basic.

9.67 The combination of a weak acid and the salt of its conjugate base is a buffer solution. The pH of a buffer is

$$\text{pH} = \text{pK}_a + \log\left(\frac{\text{proton acceptor}}{\text{proton donor}}\right) = 8.2 = 7.2 + x$$

$$x = 1.0 = \log\left(\frac{\text{proton acceptor}}{\text{proton donor}}\right) \rightarrow HPO_4^{2-}/H_2PO_4^- = 10/1$$

9.69 $62.0 \text{ mL} \times \dfrac{1\text{L}}{1000\text{ mL}} \times \dfrac{0.025\text{ mol}}{\text{L}} = 0.00155 \text{ mol HCl}$

$0.00155 \text{ mol HCl} \times \dfrac{1 \text{ mol Ca(OH)}_2}{2 \text{ mol HCl}} = 0.000775 \text{ mol Ca(OH)}_2$ when neutral

$\dfrac{0.000775 \text{ moles}}{0.025 \text{L}} = 0.031 \text{M}$

9.71 (a) HCl is a strong acid, which means that every HCl molecule dissociates into an H_3O^+ ion and a Cl⁻ ion. Thus, $[H_3O^+] = [\text{HCl}] = 0.0034$ M. It is also known that

$$[H_3O^+] \times [OH^-] = 1 \times 10^{-14} \rightarrow [OH^-] = \frac{1 \times 10^{-14}}{0.0034} = 2.94 \times 10^{-12} \text{ M}$$

Acids, Bases, and Buffers   51

$$pOH = -\log[OH^-] = 11.53$$

(b) HNO₃ is a strong acid, which means that every HNO₃ molecule dissociates into an H₃O⁺ ion and a NO₃⁻ ion. Thus, [H₃O⁺] = [HNO₃] = 0.025 M. It is also known that

$$[H_3O^+] \times [OH^-] = 1 \times 10^{-14} \rightarrow [OH^-] = \frac{1 \times 10^{-14}}{0.025} = 4.0 \times 10^{-13} \text{ M}$$

$$pOH = -\log[OH^-] = 12.40$$

(c) HBr is a strong acid, which means that every HBr molecule dissociates into an H₃O⁺ ion and a Br⁻ ion. Thus, [H₃O⁺] = [HBr] = 0.00073 M. It is also known that

$$[H_3O^+] \times [OH^-] = 1 \times 10^{-14} \rightarrow [OH^-] = \frac{1 \times 10^{-14}}{0.00073} = 1.36 \times 10^{-11} \text{ M}$$

$$pOH = -\log[OH^-] = 10.86$$

(d) HI is a strong acid, which means that every HI molecule dissociates into an H₃O⁺ ion and a I⁻ ion. Thus, [H₃O⁺] = [HI] = 0.074 M. It is also known that

$$[H_3O^+] \times [OH^-] = 1 \times 10^{-14} \rightarrow [OH^-] = \frac{1 \times 10^{-14}}{0.074} = 1.35 \times 10^{-13} \text{ M}$$

$$pOH = -\log[OH^-] = 12.87$$

9.73 Ammonium acetate is a salt that will dissociate into an NH_4^+ ion and an acetate ion, CH_3COO^-. NH_4^+ and CH_3COO^- ions can interact with water as an acid and a base, respectively. Inspection of Table 9.5 shows that the pK_a of NH4⁺ and the pK_b of CH_3COO^- are the same, 9.24. This signifies that the amount of OH⁻ and H₃O⁺ ions produced by the NH_4^+ and CH_3COO^- ions will be equal. Therefore the solution will be neutral.

9.75 A buffer is approximately equal amounts of a weak acid and a salt of its conjugate base.

9.77 A conjugate acid-base pair is a weak acid and the basic anion that results from its dissociation.

9.79 pH = $-\log[H_3O^+]$ = 1.50 → [H₃O⁺] = 0.0316 mol/liter

0.0316 mol/liter × 2 liters = 0.0632 mol of H₃O⁺.

To neutralize, 0.0632 mol of HCO₃⁻ must be added.

0.0632 mol × 84 g/mol = 5.31 g of NaCO₃ is needed.

Chapter 10

Chemical and Biological Effects of Radiation

10.1 α-Radiation is the emission of α-particles that are helium nuclei, with a charge of 2+ and a mass of 4 amu.

10.3 γ-Radiation has no mass or charge, but can penetrate very deeply into solids.

10.5 The sum of the mass numbers of the particles on the left must equal the sum of the mass numbers of the particles on the right; $240 + x = 243 + 1$, $x = 4$. Similarly, the sum of the atomic numbers of the particles on the left must equal the sum of the atomic numbers of the particles on the right; $95 + y = 97 + 0$, $y = 2$. Thus the missing component must be a particle with a mass number of 4 and an atomic number of 2. This is an α-particle, ^4_2He.

10.7 The mass number of the particle on the left must equal the sum of the mass numbers of the particles on the right; $40 = 40 + x$, $x = 0$. Similarly, the atomic number of the particles on the left must equal the sum of the atomic numbers of the particles on the right; $19 = 20 + y$, $y = -1$. Thus the missing component must be a particle with a mass number of 0 and an atomic number of -1. This is an electron, $^0_{-1}e$.

10.9 Induced radioactivity results when a nonradioactive element is bombarded with high-energy subatomic particles.

10.11 A radioactive decay series is a series of nuclear reactions that begins with an unstable nucleus and ends with the formation of a stable isotope.

10.13 $\dfrac{N}{N_0} = \left(\dfrac{1}{2}\right)^x$, where x is the number of half-lives of C-14 that have passed.

$\dfrac{N}{N_0} = 0.233$; thus $0.233 = (1/2)^x$ → $\log(0.233) = x \log(1/2)$ → $x = \log(0.233)/\log(1/2)$

  = 2.101 half-lives × 5730 years/half-life = 12,042 years.

10.15 β-Radiation passes through solids more easily than α-radiation; thus β-radiation.

10.17 As an X-ray passes through tissue, it interacts with the tissue and loses energy, and its intensity is lowered.

10.19 γ-Radiation will ionize water molecules, causing the water molecules to lose an electron and produce hydrated electrons.

Chemical and Biological Effects of Radiation 53

10.21 The secondary chemical processes have the most long-term consequences because the free radicals that are formed can undergo harmful reactions with biomolecules.

10.23 Radiation sickness is the result of a nonlethal exposure to radiation. It is characterized by nausea, lethargy, and a drop in the white blood cell count.

10.25 The intensity of radiation decreases by a factor that is inversely proportional to the square of the distance. Thus, if the technician doubles the distance between herself and the radiation, the intensity will decrease by a factor of 4. Therefore, she must move from 4 feet away to 8 feet away.

10.27 The Geiger-Müller counter measures the amount of ions produced owing to the presence of the radiation.

10.29 A rad is the amount of energy that a sample absorbs when it is exposed to radiation. This factor does not account for effects that the radiation may have on the sample.

10.31 No. α-Particles are more damaging to tissue (RBE = 10) than γ-particles (RBE = 1.0), and thus 1 rad of α-particles will be more harmful than 1 rad of γ-particles.

10.33 No. To be used for diagnosis, the radiation must be detected outside the body. α-Particles do not penetrate tissue and thus would not be detected externally.

10.35 Yes. Cobalt-60 undergoes radioactive decay to emit γ-radiation that can easily penetrate tissue to cause radiation damage to the cancerous area.

10.37 No. Positrons can be detected only when they combine with electrons to form γ-rays, which in turn are detected only when they interact with matter to form ions.

10.39 No. To be used for diagnosis, the radiation must be detected outside the body. The β-particles that result from iodine-131 do not penetrate tissue and thus remain in the body and cannot be detected externally.

10.41 A CAT scan produces images of the internal structures in the body with the use of X-rays.

10.43 30 hours = 5 half-lives. $\frac{N}{N_0} = \left(\frac{1}{2}\right)^x$, where x = 5 → $(1/2)^5 = 0.03125$; thus 0.03125×1 g = 0.03125 g = 31.25 mg remains.

10.45 $6.3 \text{ mCi} \times \frac{\text{Ci}}{1000 \text{ mCi}} \times \frac{3.7 \times 10^{10} \text{ dps}}{\text{Ci}} = 2.33 \times 10^8$ disintegrations per second

10.47 4 hours = 240 minutes. Because 110 minutes = 1 half-life, 240 minutes = 2.18 half-lives.

$$\frac{N}{N_0} = \left(\frac{1}{2}\right)^x$$, where x = 2.18 → $(1/2)^{2.18}$ = 0.221; thus 22.1% of the original sample remains after 4 hours.

10.49 Radioactivity is the name of the process in which atomic nuclei spontaneously decompose.

10.51 No. γ-radiation does not alter the atomic mass or number of an atom with which it interacts.

10.53 Any tissue that is rapidly dividing such that any damage that may result from an interaction will affect the ability of the tissue to perform its biological function.

10.55 In nuclear fission, unstable nuclei of high atomic mass are split by an interaction with a high-energy particle into nuclei with lower atomic mass. In nuclear fusion, nuclei with lower atomic mass fuse together to form new nuclei of higher atomic mass. In either case, an enormous amount of energy is released.

Chapter 11
Saturated Hydrocarbons

11.1 The number of covalent bonds formed by the atom of an element in forming a compound.

11.3 In the molecular formula, the carbons must have four bonds and the hydrogens must have one bond. The only formula that obeys this rule is formula a.

11.5 The molecular formula must be a structure that has all of the atoms attaining the correct combining power. The only formula where this is true is in b.

11.7 (a)

    C—C   and   C—H

(b)

$$\overset{\overset{\displaystyle O}{\|}}{-C-OH}$$

(c)

$$\overset{\overset{\displaystyle O}{\|}}{-C-H}$$

11.9 (a) alcohol   (b) alkene   (c) ketone   (d) aromatic

(e) ketone   (f) ether   (g) carboxylic acid   (h) amide

11.11 A carbon atom in an alkane can form four equivalent bonds by sharing electrons with other atoms; therefore the four electrons in the outer shell of carbon in the ground state must be equal. However, the four electrons in the outer shell of carbon in the ground state are not equal (one is in the s orbital; three are in p orbitals). This discrepancy is corrected by proposing that the four electrons fill four equivalent sp^3 orbitals, which are formed by combining (hybridizing) the 3 p and 1 s orbitals. Then, the four electrons can form four equivalent bonds.

11.13 The expanded structure explicitly shows each C–C and C–H bond of an alkane; therefore the answers are:

(a)   (b)

55

(c)

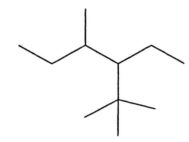

11.15 The line structural formulas show only lines to denote the C–C bonds. Bends are put in to clearly denote where atoms exist. Therefore, the answers are:

(a)

(b)

(c)

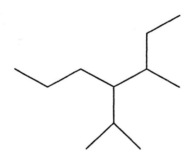

(d)

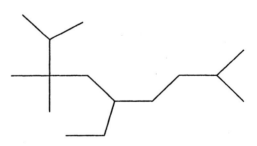

11.17 (a) Both compounds have a molecular formula of C_5H_{12}; therefore they are either the same compound or constitutional isomers. Drawing the expanded structural formula of the first compound reveals that they are the same compound.

(b) Both compounds have a molecular formula of C_6H_{14}; therefore they are either the same compound or constitutional isomers. Drawing the expanded structural formula of the first compound shows that they are not the same compound; therefore they are constitutional isomers.

(c) Both compounds have a molecular formula of C_6H_{14}; therefore they are either the same compound or constitutional isomers. Drawing the expanded structural formula of the second compound reveals that they are not the same compound; therefore they are constitutional isomers.

11.19

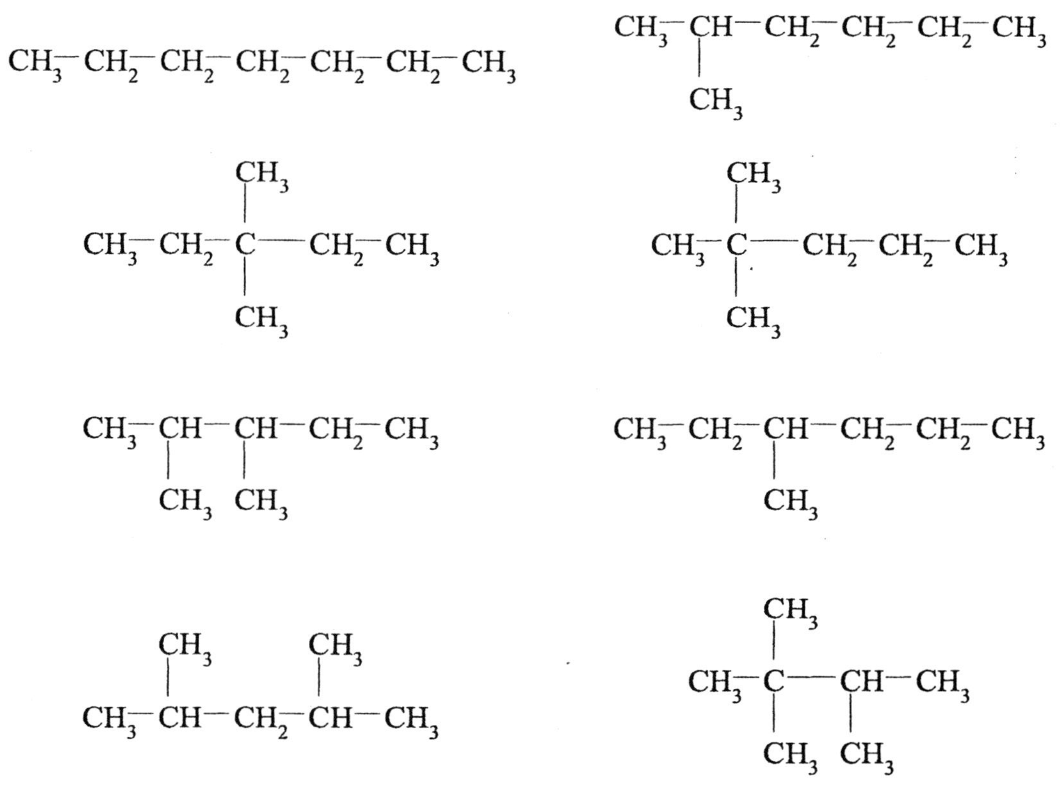

$CH_3-CH_2-CH_2-CH_2-CH_2-CH_2-CH_3$

$CH_3-CH(CH_3)-CH_2-CH_2-CH_2-CH_3$

$CH_3-CH_2-C(CH_3)(CH_3)-CH_2-CH_3$

$CH_3-C(CH_3)(CH_3)-CH_2-CH_2-CH_3$

$CH_3-CH(CH_3)-CH(CH_3)-CH_2-CH_3$

$CH_3-CH_2-CH(CH_3)-CH_2-CH_2-CH_3$

$CH_3-CH(CH_3)-CH_2-CH(CH_3)-CH_3$

$CH_3-C(CH_3)(CH_3)-CH(CH_3)-CH_3$

11.21 Method for naming alkanes:

(i) Find longest carbon chain. Form base name accordingly.

(ii) Find and name groups that are attached to this primary chain.

(iii) Put names of these groups in alphabetical order at beginning of name. Include numbers to denote the carbon(s) to which the groups are attached. Number the carbons such that the smallest numbers are used.

(a) (i) Four carbons = butane

(ii) One CH_3 group = methyl attached

(iii) 2-Methylbutane

(b) (i) Five carbons = pentane

(ii) One CH_3 group = methyl attached

(iii) 3-Methylpentane

(c) (i) Six carbons = hexane

(ii) Three CH₃ groups = methyls attached

(iii) 2,2,4-Trimethylhexane

(d) (i) Six carbons = hexane

(ii) One CH₃ group = methyl and one CH₃CHCH₃ = isopropyl group

(iii) 3-Isopropyl-4-methylhexane

This compound is interesting in that, if pictured as the below, it can also be (correctly) named as follows:

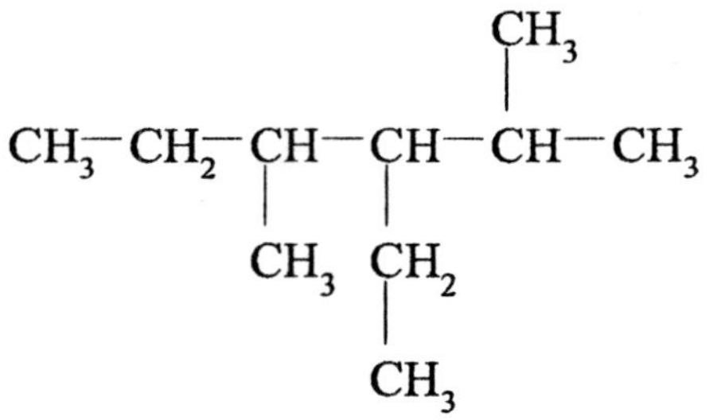

$$CH_3-CH_2-\underset{\underset{CH_3}{|}}{CH}-\underset{\underset{\underset{\underset{CH_3}{|}}{CH_2}}{|}}{CH}-\underset{\overset{\overset{CH_3}{|}}{|}}{CH}-CH_3$$

(i) Six carbons = hexane

(ii) Two CH₃ groups = methyls and one CH₂CH₃ = ethyl group

(iii) 3-Ethyl-2,4-dimethylhexane

This exemplifies the fact that the IUPAC naming rules are guidelines.

11.23 Method for drawing structures from alkane names:

(i) Draw longest carbon chain from the base name.

(ii) Add side groups on appropriate carbons along chain.

(iii) Fill in hydrogens such that each carbon has four bonds.

(a) (i) Butane = 4 carbons:

C–C–C–C

(ii) 2-Methyl = CH₃ group on the second carbon.

(iii)

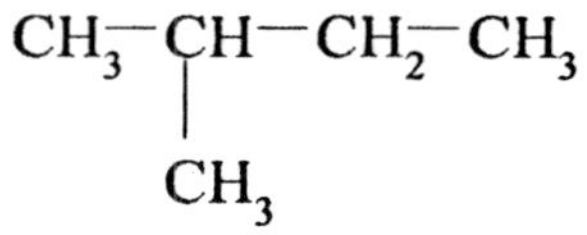

$$CH_3-\underset{\underset{CH_3}{|}}{CH}-CH_2-CH_3$$

(b) (i) Hexane = six carbons:

C–C–C–C–C–C

(ii) 2,2,4-Trimethyl = CH₃ on C-2, C-2, and C-4.

(iii)

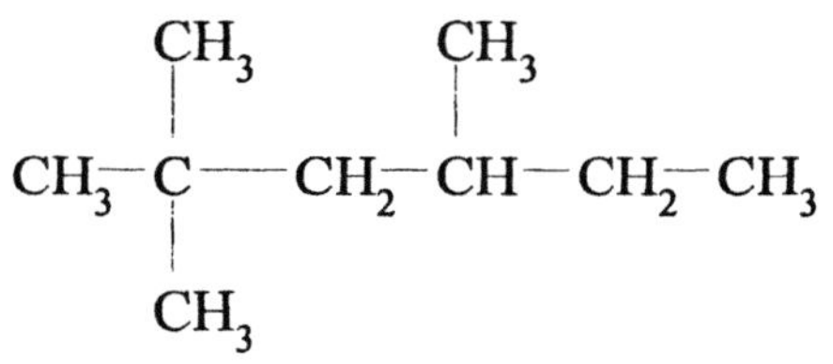

$$CH_3-\underset{\underset{CH_3}{|}}{\overset{\overset{CH_3}{|}}{C}}-CH_2-\underset{\underset{}{}}{\overset{\overset{CH_3}{|}}{CH}}-CH_2-CH_3$$

(c) (i) Hexane = six carbons:

C–C–C–C–C–C

(ii) 2,3-Dimethyl = CH_3 on C-2 and C-3; 3-isopropyl = CH_3CHCH_3 on C-3

(iii)

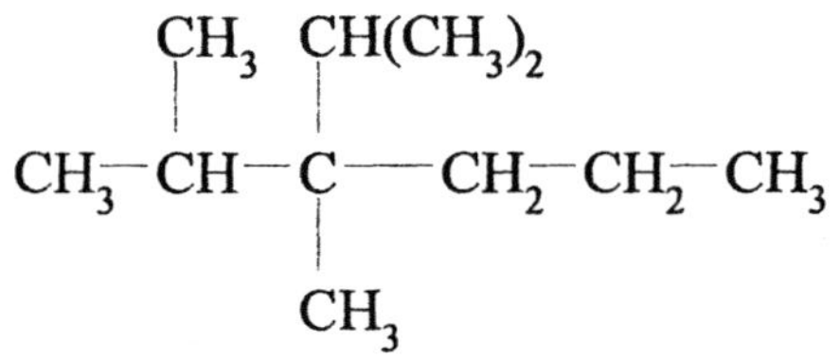

$$CH_3-CH-\underset{\underset{CH_3}{|}}{\overset{\overset{CH_3\ \ CH(CH_3)_2}{|\ \ \ \ \ |}}{C}}-CH_2-CH_2-CH_3$$

(d) (i) Octane = eight carbons:

C–C–C–C–C–C–C–C

(ii) 2,2-Dimethyl = CH_3 on C-2 and C-2; 4-isopropyl = CH_3CHCH_3 on C-4; 4-ethyl = CH_2CH_3 on C-4.

(iii)

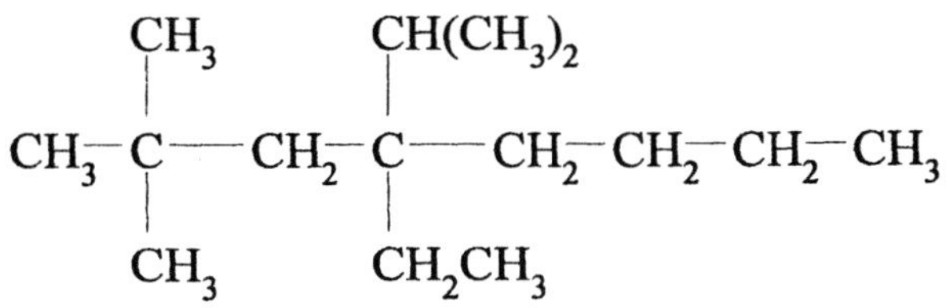

$$CH_3-\underset{\underset{CH_3}{|}}{\overset{\overset{CH_3}{|}}{C}}-CH_2-\underset{\underset{CH_2CH_3}{|}}{\overset{\overset{CH(CH_3)_2}{|}}{C}}-CH_2-CH_2-CH_2-CH_3$$

(e) (i) Heptane = 7 carbons:

C–C–C–C–C–C–C

(ii) 2-Methyl = CH_3 on C-2; 4-*s*-butyl = $CH_3CH_2CHCH_3$ on C-4.

(iii)

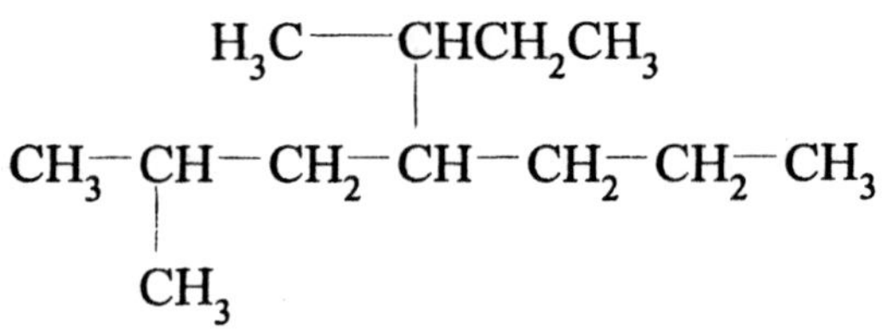

$$H_3C-CHCH_2CH_3$$
$$\hspace{2.5em}|$$
$$CH_3-CH-CH_2-CH-CH_2-CH_2-CH_3$$
$$\hspace{1em}|$$
$$\hspace{1em}CH_3$$

11.25 (a) Octane = 8 carbons; thus:

(b) Cyclohexene = six carbons:

Choose a carbon for one of the methyl groups, label that carbon in the ring C-1.

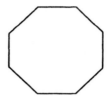
$\hspace{4em}$CH$_3$

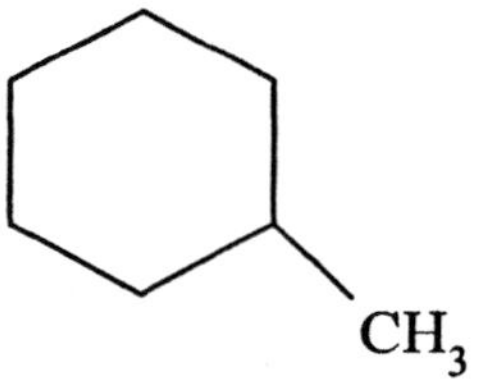

Place other groups (methyl and isopropyl) on the correct carbons by counting clockwise around ring from C-1.

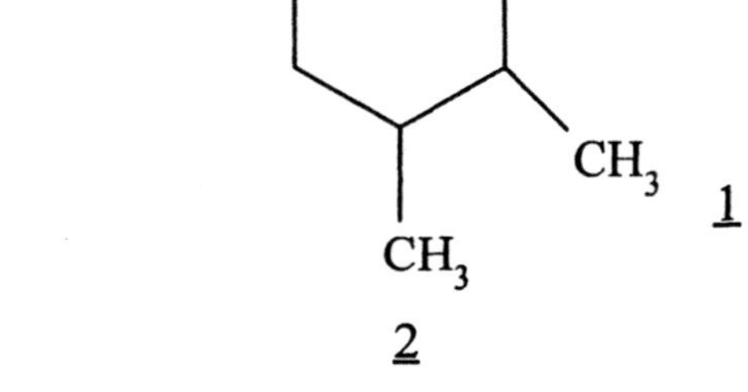

(c) Cyclohexene = 6 carbons:

Choose a carbon for *t*-butyl group; label that carbon in the ring as C-1.

Place other group (ethyl) on the correct carbon (C-3) by counting clockwise around ring from C-1.

11.27 A 1° (primary) carbon is connected to one other carbon.

A 2° (secondary) carbon is connected to two other carbons.

A 3° (tertiary) carbon is connected to three other carbons.

A 4° (quarternary) carbon is connected to four other carbons.

(a)

(b)

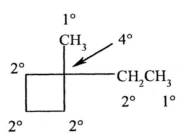

(c)

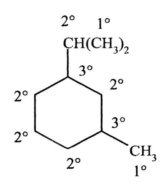

11.29 (a) One methyl (CH$_3$) group cannot be part of a ring; therefore it must be connected to the ring. This leaves C$_5$H$_9$ to form the (cyclopentane) ring. The compound is methylcyclopentane.

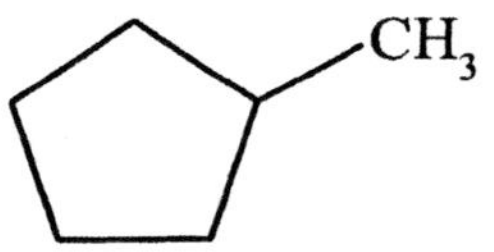

(b) An ethyl group = C$_2$H$_5$, which cannot be part of the ring structure; thus C$_4$H$_7$ remains to form the ring. Therefore the ring is cyclobutane. The compound is thus ethylcyclobutane.

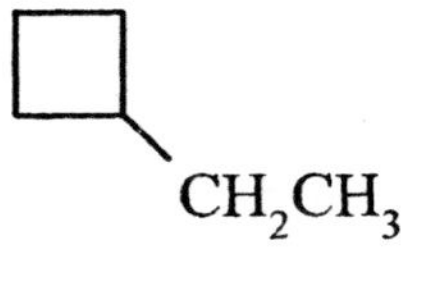

11.31 To exist as cis-trans isomers in cycloalkanes, there must exist two carbons within the ring which each contains 2 different substituents.

(a) No cis-trans isomerism, because only one carbon has two different substituents on it.

(b) Both C-1 and C-2 have two different substituents; therefore this compound does exhibit cis-trans isomers:

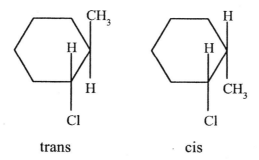

(c) Both C-1 and C-3 have two different substituents; therefore this compound does exhibit cis-trans isomers:

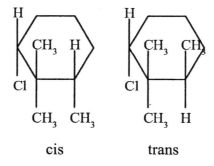

11.33 (a) Hexane has the higher boiling point because it has the higher molecular mass and; therefore, has greater secondary forces.

(b) Cyclohexane has the higher boiling point because it has the higher molecular mass and; therefore, has greater secondary forces.

(c) Methylcyclohexane has the higher boiling point because it has the higher molecular mass and; therefore, has greater secondary forces.

(d) Hexane has the higher boiling point because it is not branched and; therefore, has greater secondary forces.

(e) Cyclopentane has the higher boiling point because it is cyclic, packs tightly, and, therefore, has greater secondary forces.

11.35 Alkanes (and cycloalkanes) are inert substances that do not undergo many reactions. The only two common reactions that they undergo are (i) combustion with oxygen and (ii) halogenation = substitution of a halogen (F, Cl, Br, etc.) for a hydrogen in the presence of UV or heat.

Thus, there is no reaction for a, b, c, and d.

(e) This halogenation reaction will proceed in which a Cl will replace one of the hydrogens on the butane to form:

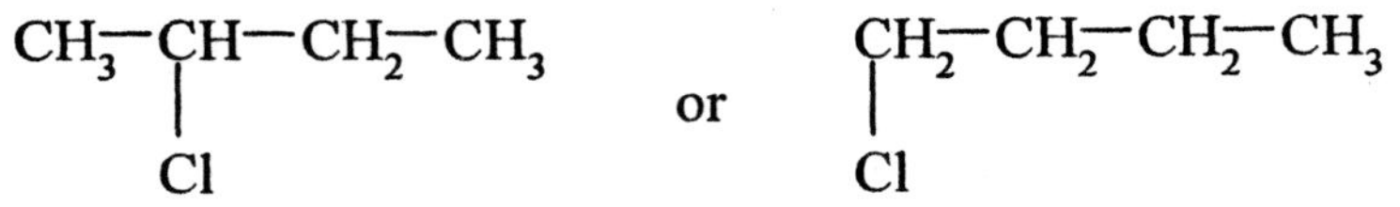

$$CH_3-CH-CH_2-CH_3 \quad \text{or} \quad CH_2-CH_2-CH_2-CH_3$$
$$\quad \;\;|\qquad\qquad\qquad\qquad\qquad\;\; |$$
$$\quad \;\;Cl\qquad\qquad\qquad\qquad\qquad\;\; Cl$$

11.37 Complete combustion of an alkane, C_nH_m, is given by the following equation:

$$C_nH_m + z\, O_2 \rightarrow n\, CO_2 + (m/2)\, H_2O$$

where $z = (n + m/2)$

(a) Pentane = C_5H_{12}; thus n = 5, m = 12, z = 8.

$$C_5H_{12} + 8\, O_2 \rightarrow 5\, CO_2 + 6\, H_2O$$

(b) 2-Methylbutane = C_5H_{12}; thus n = 5, m = 12, z = 8.

$$C_5H_{12} + 8\, O_2 \rightarrow 5\, CO_2 + 6\, H_2O$$

(c) Cyclohexane = C_6H_{12}; thus n = 6, m = 12, z = 9.

$$C_6H_{12} + 9\, O_2 \rightarrow 6\, CO_2 + 6\, H_2O$$

(d) Methylcyclopentane = C_6H_{12}; thus n = 6, m = 12, z = 9.

$$C_6H_{12} + 9\, O_2 \rightarrow 6\, CO_2 + 6\, H_2O$$

11.39 One carbon can bond to one or more other carbon atoms in many different patterns.

11.41

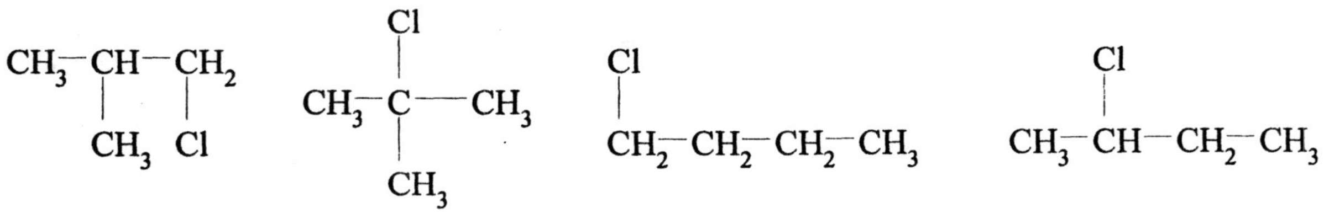

$$CH_3-CH-CH_2 \qquad\qquad CH_3-\underset{\underset{CH_3}{|}}{\overset{\overset{Cl}{|}}{C}}-CH_3 \qquad\qquad \underset{CH_2-CH_2-CH_2-CH_3}{\overset{\overset{Cl}{|}}{}} \qquad\qquad \underset{CH_3-CH-CH_2-CH_3}{\overset{\overset{Cl}{|}}{}}$$
$$\;\;|\qquad |$$
$$CH_3\;\; Cl$$

11.43 (a) Because the first compound has one chloro group, whereas the second compound has two Cl, these two structures must be different compounds that are not isomers.

(b) Because the first compound is a cycloalkane and the second compound is not, these two structures must be different compounds that are not isomers.

(c) Both compounds have a molecular formula of C_9H_{18}; therefore they are either the same compound or isomers. Because the first compound has a five-membered ring, whereas the second compound has a six-membered ring, the two compounds must be constitutional isomers.

(d) Both compounds have a molecular formula of C_9H_{18}; therefore they are either the same compound or isomers. Because the first compound has both a methyl and an ethyl group connected to the ring, whereas the second compound has a single propyl group connected to the ring, they must be constitutional isomers.

(e) The molecular formula of the first compound is C_9H_{18}, whereas that of the second compound is C_9H_{20}; these two structures must be different compounds that are not isomers.

(f) Both compounds have a molecular formula of $C_4H_{11}Cl$; therefore they are either the same compound or isomers. Because the first compound has a tertiary carbon, whereas the second compound does not, the two compounds must be constitutional isomers.

(g) Both compounds have a molecular formula of $C_4H_{11}Cl$; therefore they are either the same compound or isomers. Because the first compound has a quarternary carbon, whereas the second does not, the two compounds must be constitutional isomers.

(h) Both compounds have a molecular formula of $C_6H_{10}Cl_2$; therefore they are either the same compound or isomers. Because the first compound has both chloro groups on the same side of the ring, whereas the second compound has them on opposite sides, they must be cis-trans isomers.

(i) Both compounds have a molecular formula of $C_6H_{10}Cl_2$; therefore they are either the same compound or isomers. Because the first compound has the chloro groups on C-1 and C-3, whereas the second compound has them on C-1 and C-2, they must be constitutional isomers.

(j) Both compounds have a molecular formula of $C_4H_8Cl_2$; therefore they are either the same compound or isomers. Expansion of the first compound shows that the two structures are the same compound.

11.45 Cyclic alkanes have the general formula C_nH_{2n}, whereas acyclic alkanes have the formula C_nH_{2n+2}. $C_5H_{10} = C_nH_{2n}$ when n = 5; therefore this compound must be a cyclic alkane.

11.47 Count the carbons, hydrogens and other atoms to get:

(a) C_6H_{14} (b) C_8H_{18} (c) $C_6H_{11}Cl$ (d) $C_6H_{11}Cl$

(e) C_9H_{20} (f) C_6H_{12}

11.49 (a) Complete combustion of an alkane, C_nH_m, is given by the following equation:

$$C_nH_m + z\, O_2 \rightarrow n\, CO_2 + (m/2)\, H_2O$$

where $z = (n + m/2)$

$$C_6H_{14} + 19/2\, O_2 \rightarrow 6\, CO_2 + 7\, H_2O \equiv 2\, C_6H_{14} + 19\, O_2 \rightarrow 12\, CO_2 + 14\, H_2O$$

(b) In a monobromination reaction, the bromine can replace any hydrogen to form different isomers:

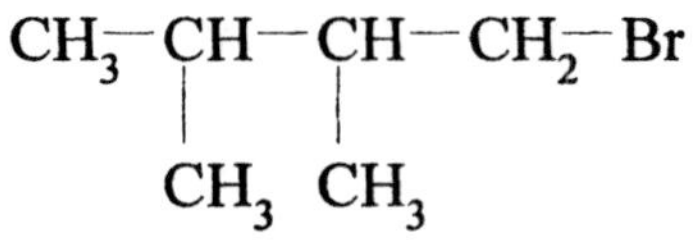

$$\underset{\underset{CH_3\ \ CH_3}{|\ \ \ \ \ |}}{CH_3-CH-CH-CH_2-Br} \quad \text{and} \quad \underset{\underset{Br}{|}}{\overset{\overset{CH_3\ \ CH_3}{|\ \ \ \ \ |}}{CH_3-C-CH-CH_3}}$$

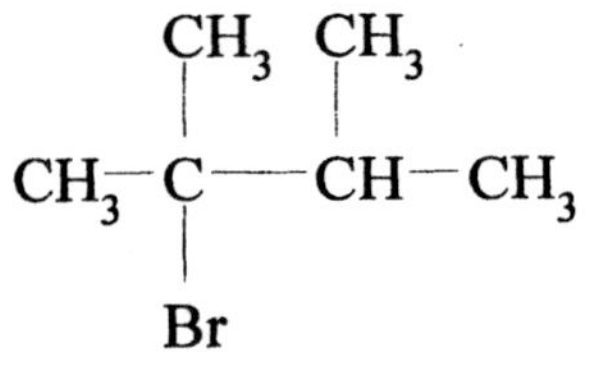

11.51 $CH_3-CH_2-CH_2-CHCl_2$ and $CH_3-CH_2-CCl_2-CH_3$ and $CH_2Cl-CHCl-CH_2-CH_3$ and $CH_2Cl-CH_2-CHCl-CH_3$ and $CH_2Cl-CH_2-CH_2-CH_2Cl$

11.53 The alkanes in oil possess only the weak London attractive forces, whereas water contains the much stronger hydrogen-bonding forces. Alkane molecules are much more weakly attracted to each other than are water molecules, which results in a lower density for oil.

11.55 Chemical reactions are used to convert alkanes into members of other families.

11.57 The molecules of lipstick and petroleum jelly attract each other because they are both hydrocarbons with the same type of secondary force (London). Therefore petroleum jelly dissolves the lipstick, whereas H_2O does not.

11.59 $160.43\text{ g } CH_4 \times 1 \text{ mole} / 16.04 \text{ g} = 10 \text{ moles of } CH_4$

$10 \text{ mole } CH_4 \times 803 \text{ kJ} / 1 \text{ mole } CH_4 = 8{,}030 \text{ kJ}$

Chapter 12

Unsaturated Hydrocarbons

12.1 An alkane has the formula C_nH_{2n+2}, whereas alkenes or cycloalkanes have the formula C_nH_{2n}. If n = 5, then the formula of an alkane will be C_5H_{12}.

12.3 An alkane has the formula C_nH_{2n+2}, whereas alkenes or cycloalkanes have the formula C_nH_{2n}. A single hydrogen in these formula can also be exchanged with a halogen (Group VII) atom. If a molecular formula of a compound does not follow these guidelines, a compound in which each of the carbon atoms has four bonds cannot be created. Therefore, (a) incorrect; (b) correct; (c) correct.

12.5 Counting the number of carbons and hydrogens gives (a) C_5H_{10} and (b) C_7H_{12}. In part b, note that the carbon that has a CH_3 connected to it does not have a hydrogen.

12.7 sp^2 hybridization means that the carbon atom has three equivalent orbitals and one remaining p orbital. This hybridization is used in the formation of carbon–carbon double bonds and thus those carbons that are part of a double bond have sp^2 hybridization.

Therefore, carbons 3, 4, 8, and 9 have sp^2 hybridization.

12.9

$CH_2=CH-CH_2-CH_3$ $CH_3-CH=CH-CH_3$
                                                       (cis and trans)

$CH_2=C(CH_3)-CH_3$ ☐   △—CH_3

12.11 (a) C–C–C–C–C is the skeleton. One of the bonds between carbons must be a double bond; therefore possibilities include:

C=C–C–C–C   and   C–C=C–C–C

Filling in the hydrogens gives:

$CH_2=CH-CH_2-CH_2-CH_3$ and $CH_3-CH=CH-CH_2-CH_3$ (cis and trans)

68 Chapter 12

(b) Methyl groups (CH₃) must be on the end of a chain and cannot be part of a double bond. Therefore, the skeleton must be:

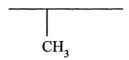

And the carbon–carbon double bond must be between the second and third carbons of the main chain. Putting this double bond in and filling in the hydrogens gives:

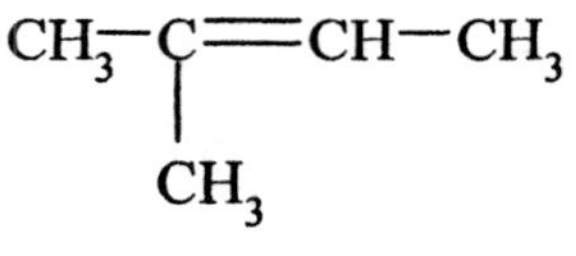

$$CH_3-C=CH-CH_3$$
$$\quad\quad |$$
$$\quad\quad CH_3$$

(c) A methyl group (CH₃) must be bonded to the ring. If there exist three CH₂ groups, then the other carbon must be bonded to only one hydrogen. This C–H bond must be at the carbon where a side group is attached to a ring. Thus, the answer must be:

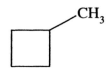

(d) An isopropyl group is:

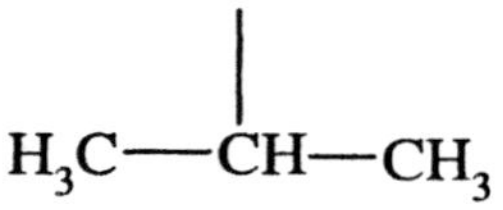

$$\quad\quad\quad |$$
$$H_3C-CH-CH_3$$

The other two carbons must be attached to the isopropyl group and therefore the double bond must be between these two carbons. Thus, the final correct structure is:

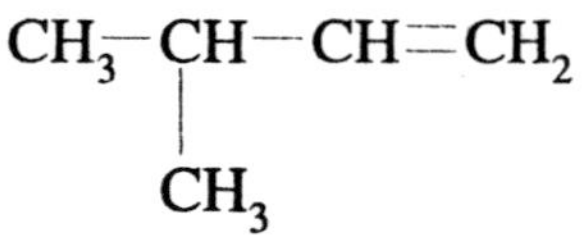

$$CH_3-CH-CH=CH_2$$
$$\quad\quad |$$
$$\quad\quad CH_3$$

12.13 (a) Because the first structure has one chlorine atom and the second structure has two chlorine atoms, they must be different structures.

(b) Because the first structure has four carbon atoms and the second structure has five carbon atoms, they must be different structures.

(c) Because both structures have four carbons and eight hydrogens and yet cannot be superimposed, they must be constitutional isomers.

12.15 (a) 2-Methyl-1-butene: Butene means that the largest chain has four carbons in it and a double bond. The number 1 before the butene means that the first carbon in the chain is part of the double bond. The 2-methyl means that there is also a methyl group on the second carbon. Thus the correct structure is:

$$CH_2=C-CH_2-CH_3$$
$$|$$
$$CH_3$$

(b) 3-Ethyl-2-pentene: Pentene means that the largest chain has five carbons in it and a double bond. The number 2 before the pentene means that the second carbon in the chain is part of the double bond. The 3-ethyl means that there is an ethyl group on the third carbon. Thus the correct structure is:

$$CH_2-CH_3$$
$$|$$
$$CH_3-CH=C-CH_2-CH_3$$

(c) 4-Isopropyl-2,6-dimethyl-2-heptene: Heptene means that the largest chain has seven carbons in it and a double bond. The number 2 before the heptene means that the second carbon in the chain is part of the double bond. The 4-isopropyl means that there is an isopropyl group on the fourth carbon. 2,6-Dimethyl means that there are two methyl groups attached to the main chain at the second and sixth carbons. Thus the correct structure is:

$$CH_3 CH_3$$
$$| |$$
$$CH_3-C=CH-CH-CH_2-CH-CH_3$$
$$|$$
$$CH_3-CH-CH_3$$

(d) 1,3-Dimethylcyclohexene: Cyclohexene means that there is a six-membered ring with a double bond in it. This double bond defines the 1 carbon. The 1,3-dimethyl means that there are two methyl groups attached to the ring at the first and third carbons. Thus the correct structure is:

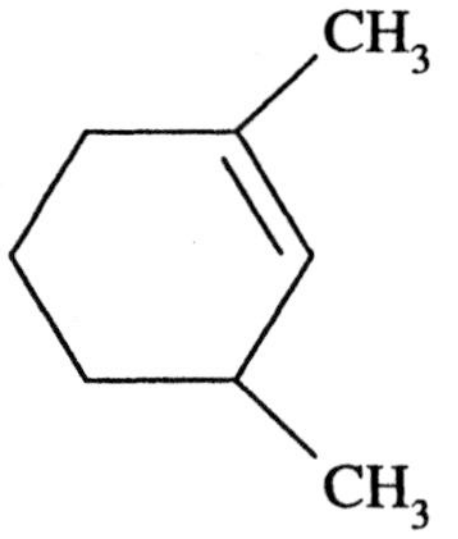

(e) 3-*t*-Butyl-2,4-dimethyl-1-pentene: Pentene means that the largest chain has five carbons in it and a double bond. The 1 before the pentene means that the first carbon in the chain is part of the double bond. The 3-*t*-butyl means that there is also a tert-butyl group on the third carbon. 2,4-Dimethyl means that there are two methyl groups attached to the main chain at the second and fourth carbons. Thus the correct structure is:

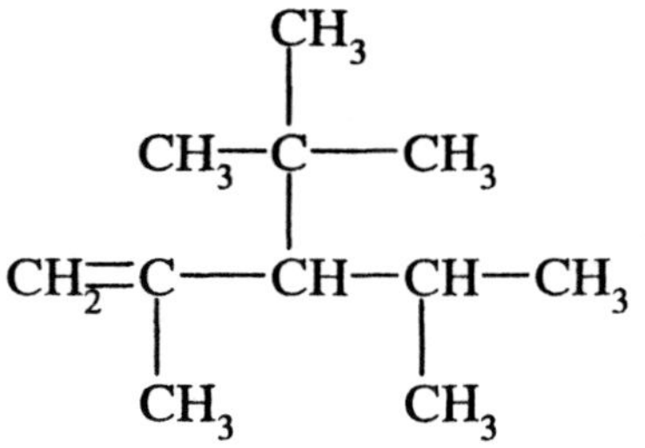

(f) 5-Methyl-1,4-hexadiene: Hexadiene means that the largest chain has six carbons in it and two double bonds. The 1,4 before the hexadiene means that the first carbon and the fourth carbon in the chain are part of double bonds. The 5-methyl means that there is a methyl group on the fifth carbon. Thus the correct structure is:

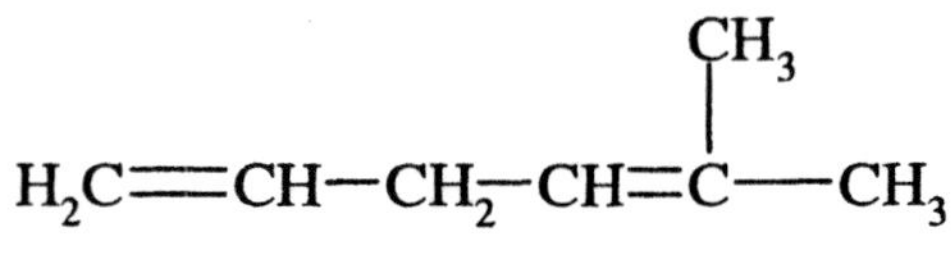

12.17 (a) The first carbon of the double bond has two methyl groups attached to it; therefore this molecule cannot have cis-trans isomers.

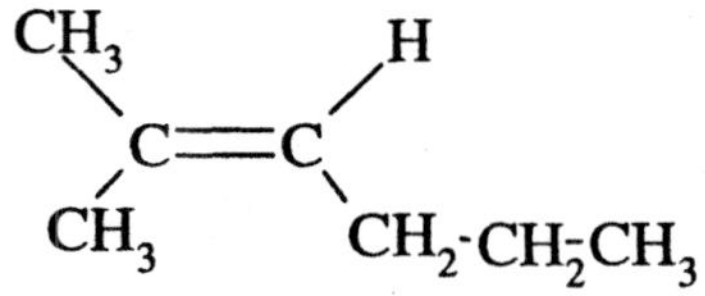

(b) 3-Methyl-2-hexene is:

$$\underset{H}{\overset{CH_3}{\diagdown}}C=C\underset{CH_3}{\overset{CH_2-CH_2-CH_3}{\diagup}} \qquad \underset{CH_3}{\overset{H}{\diagdown}}C=C\underset{CH_3}{\overset{CH_2-CH_2-CH_3}{\diagup}}$$

Because both carbons of the double bond have two different groups, this compound can form cis-trans isomers as shown above.

(c) 4-Methyl-2-hexene is:

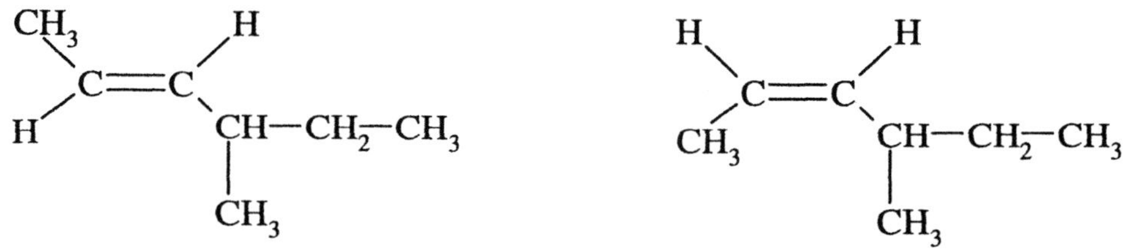

$$\underset{H}{\overset{CH_3}{\diagdown}}C=C\underset{\underset{CH_3}{|}}{\overset{H}{\diagup}}CH-CH_2-CH_3 \qquad \underset{CH_3}{\overset{H}{\diagdown}}C=C\underset{\underset{CH_3}{|}}{\overset{H}{\diagup}}CH-CH_2-CH_3$$

Because both carbons of the double bond have two different groups, this compound can form cis-trans isomers as shown above.

(d) 2-Methyl-1-hexene is:

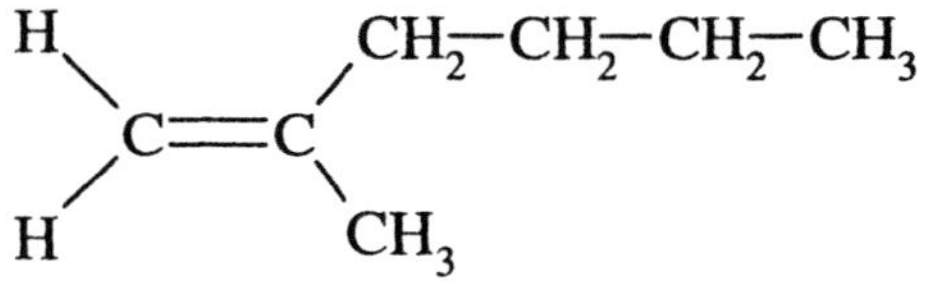

$$\underset{H}{\overset{H}{\diagdown}}C=C\underset{CH_3}{\overset{CH_2-CH_2-CH_2-CH_3}{\diagup}}$$

The first carbon of the double bond has two hydrogens attached to it; therefore this molecule cannot have cis-trans isomers.

(e) 1,2-Dimethylcyclopentene is:

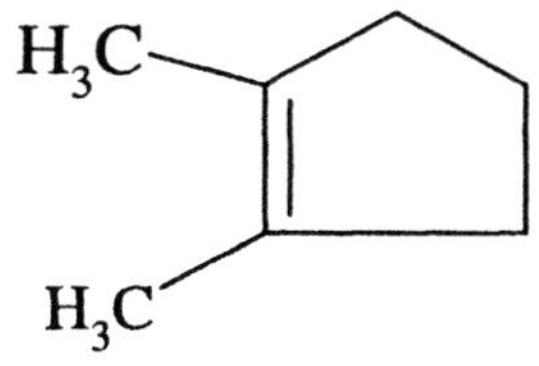

The two carbons that have methyl groups do not have another bond outside the ring; therefore this molecule cannot have cis-trans isomers.

(f) 1,2-Dimethylcyclopentane is:

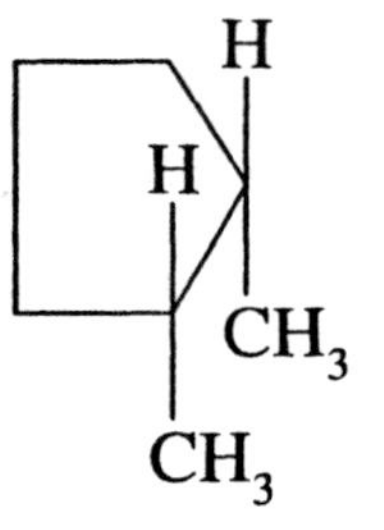

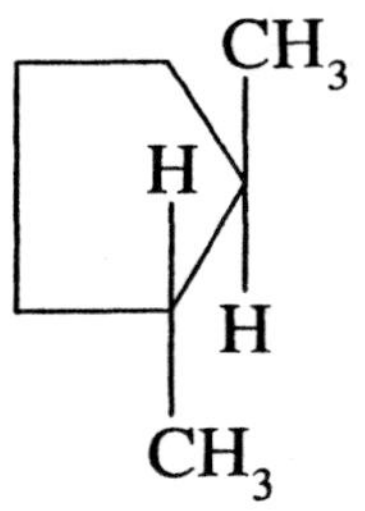

This compound can form cis-trans isomers as shown above.

12.19 (a) Both compounds have a formula of $C_4H_6Cl_2$; therefore they are isomers or the same compound. Because there is no free rotation about a double bond, the two compounds cannot be superimposed on each other; therefore they are not the same compound. Inspection shows that they are cis-trans isomers of each other.

(b) If the first compound is rotated about the C=C double bond, it will become the second compound; therefore these two structures are the same compound.

(c) These two compounds cannot be superimposed on each other; they are not the same compound. They are also not cis-trans isomers, because the compound on the right cannot have cis-trans isomers. They do have the same chemical formula, and therefore they must be constitutional isomers.

(d) The first compound has four carbons, whereas the second compound has five carbons; therefore these two compounds must be different and not isomers.

12.21 (a) Bromine adds to double bonds at both carbons; therefore the result of the reaction is in both the cis and the trans form:

(b) Water adds to the double bond in the presence of an acid catalyst to form:

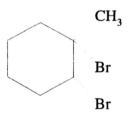

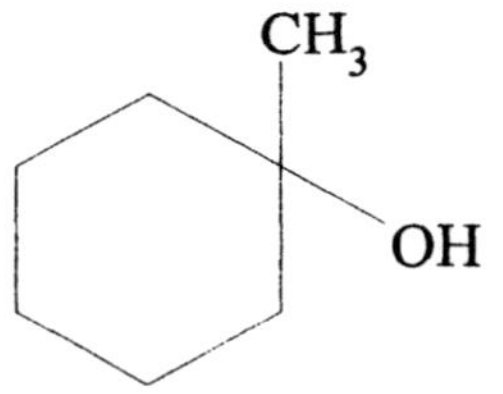

(c) Hydrogen adds to the double bond in the presence of nickel to form:

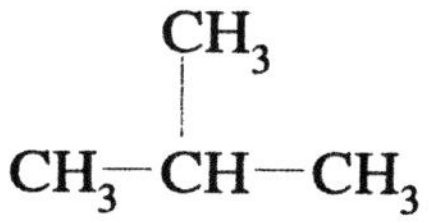

$$CH_3-\underset{\underset{CH_3}{|}}{CH}-CH_3$$

(d) HCl adds to a double bond. Note that Markovnikov's rule means that hydrogen adds to the primary carbon (the one with more H's in the reactant structure).

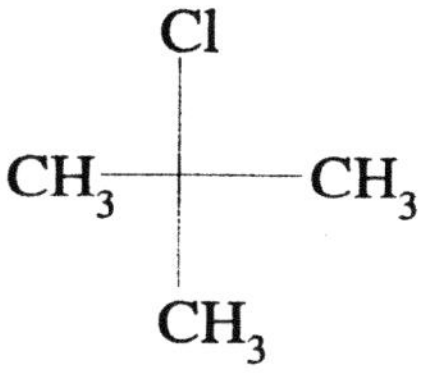

$$CH_3-\underset{\underset{CH_3}{|}}{\overset{\overset{Cl}{|}}{C}}-CH_3$$

(e) Chlorine adds to the double bond to form:

[cyclohexane ring with Cl and Cl on adjacent carbons]

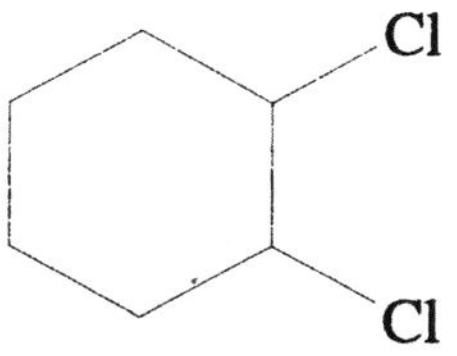

(f) HCl adds to the double bond to form:

$$CH_3-CH_2-\underset{\underset{}{}}{\overset{\overset{Cl}{|}}{CH}}-CH_3$$

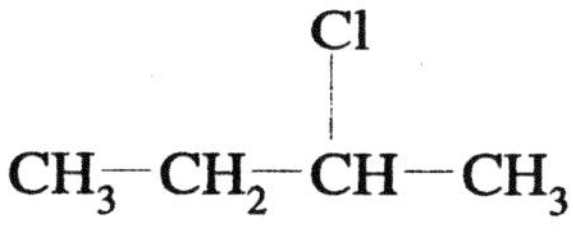

(g) HBr adds to the double bond to form:

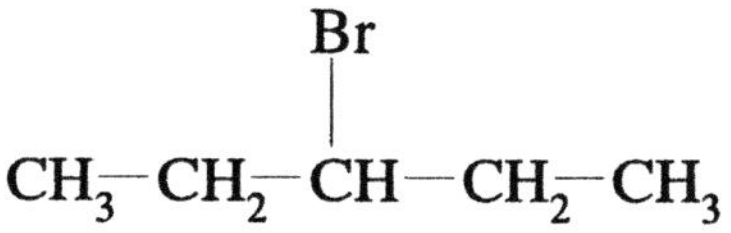

$$CH_3-CH_2-\underset{\underset{}{}}{\overset{\overset{Br}{|}}{CH}}-CH_2-CH_3 \quad \text{and} \quad CH_3-\underset{\underset{Br}{|}}{CH}-CH_2-CH_2-CH_3$$

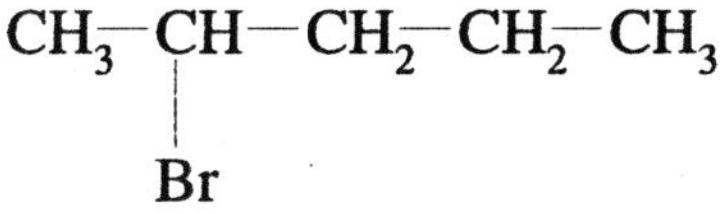

Both compounds are formed because the two carbons that form the double bond are equivalent.

(h) Br₂ adds to the double bond to form:

$$CH_3-\overset{\overset{Br}{|}}{CH}-\overset{\overset{Br}{|}}{CH}-CH_2-CH_3$$

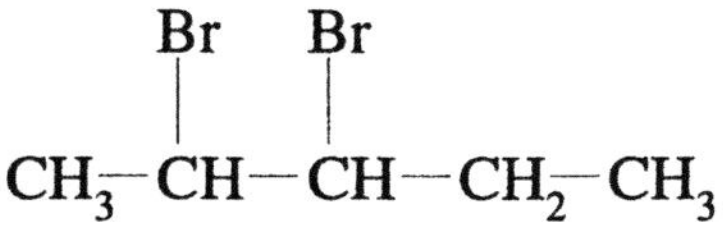

(i) Cl₂ adds to the double bond to form:

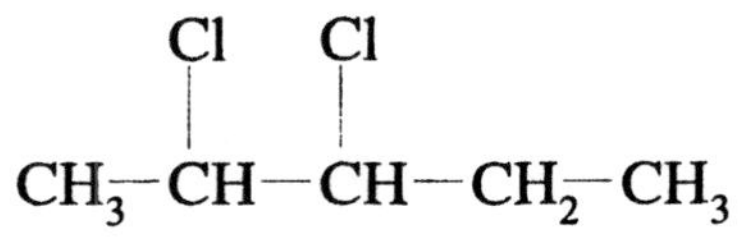

$$CH_3-\underset{\underset{Cl}{|}}{CH}-\underset{\underset{Cl}{|}}{CH}-CH_2-CH_3$$

(j) In the presence of an acid, water will add to the double bond. Because each of the two carbons that form the double bond has a single hydrogen, they are equivalent, and therefore the following two compounds form:

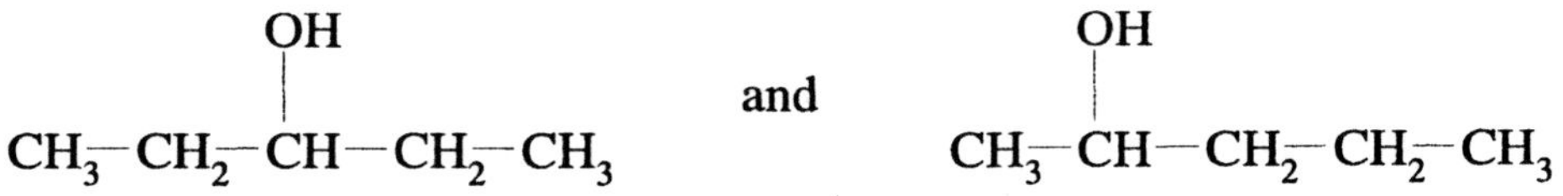

$$CH_3-CH_2-\underset{\underset{OH}{|}}{CH}-CH_2-CH_3 \quad \text{and} \quad CH_3-\underset{\underset{OH}{|}}{CH}-CH_2-CH_2-CH_3$$

(k) Hydrogen adds to the double bond in the presence of nickel to form:

$$CH_3-CH_2-CH_2-CH_2-CH_3$$

12.23 An alkane is not very reactive, whereas an alkene will undergo addition reactions under the proper conditions. One possibility is the addition of bromine (Br₂) to the double bond. The completion of this reaction can be verified by the loss of the red color of Br₂ when the reaction proceeds. Because the red color of bromine does not disappear in this process, there must be no reaction with Br₂. Therefore, the compound must not be an alkene; it must be an alkane, either hexane or methylcyclopentane.

12.25 During a polymerization, the double bond of an alkene is opened to form two more single bonds on the carbons, which grow into polymer chains by adding to other double bonds.

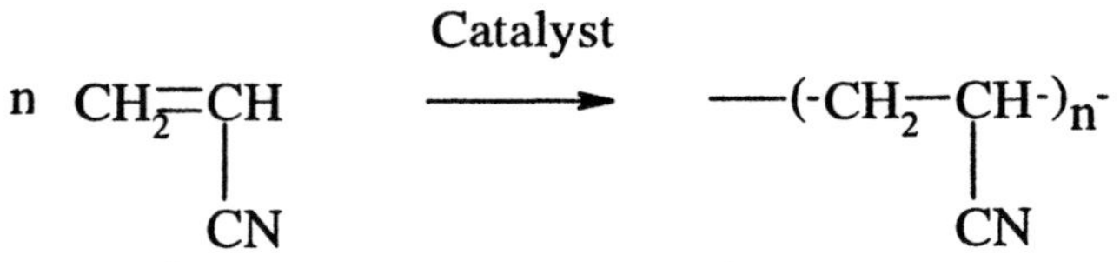

$$n\ CH_2=\underset{\underset{CN}{|}}{CH} \xrightarrow{\text{Catalyst}} -(-CH_2-\underset{\underset{CN}{|}}{CH}-)_n-$$

12.27 For the combustion of an alkene, all of the carbon of the alkene goes to carbon dioxide, whereas all of the hydrogen of the alkene goes to water. Therefore, the balanced equation must be:

$$C_nH_{2n} + 3n/2\ O_2 \rightarrow n\ CO_2 + n\ H_2O$$

For 3-hexene, n = 6; therefore the equation is:

$$C_6H_{12} + 9\ O_2 \rightarrow 6\ CO_2 + 6\ H_2O$$

12.29 An alkane is not very reactive, whereas an alkene will undergo reactions under the proper conditions. One possibility is the reaction of chromate to the double bond. The completion of this reaction can be verified by the loss of the orange color of chromate when the reaction proceeds. Because the orange color does not disappear in this process, the compound must not be an alkene. It therefore must be an alkane, either pentane or cyclopentane.

12.31 Pentyne means that there are five carbons in the main chain and a triple bond. The number 1 before the pentyne signifies that the triple bond is on the first carbon. The 3,4-dimethyl means that there are two methyl groups connected to the main chain at the third and fourth carbons. Thus,

$$CH\equiv C-\underset{CH_3}{\underset{|}{CH}}-\underset{CH_3}{\underset{|}{CH}}-CH_3$$

12.33 Aromatic compounds have three double bonds alternating with three single bonds in a six-membered ring. Aromatic compounds are very stable and do not undergo addition reactions easily. Substitution reactions (replace a hydrogen with another functional group) take place without loss of aromatic character.

12.35 (a) If the first compound is rotated about the center of the aromatic ring, it will become the second compound; therefore these two are the same compound.

(b) Both compounds have a formula of C_7H_7Cl; therefore they are isomers or the same compound. The two compounds cannot be superimposed on each other; therefore they are not the same compound. Inspection shows that they are constitutional isomers of each other.

(c) If the first compound is flipped about the axis defined by the carbon between the CH_3 and Cl groups, it will become the second compound; therefore these two are the same compound.

12.37 (a) An aromatic ring with a CH_3 group on it is toluene. This CH_3 group defines the 1-position in the ring. Therefore the chloro group is in the 3-position, or meta to the methyl group. Thus, 3-chlorotoluene or m-chlorotoluene is the correct IUPAC name.

(b) An aromatic ring with a CH_3 group on it is toluene. This CH_3 group defines the 1-position in the ring. Therefore the bromo group is in the 4-position, or para to the methyl group. Thus, 4-bromotoluene or p-bromotoluene is the correct IUPAC name.

(c) An aromatic ring with a $CH_2C_6H_5$ group is the benzyl group. The chloro group is three carbons away from it, or in the meta position. Thus, 1-benzyl-3-chlorobenzene or m-benzylchlorobenzene is the correct IUPAC name.

(d) An aromatic ring with an OH group on it is phenol. This OH group defines the 1-position in the ring. Therefore the bromo group is in the 2-position, or ortho to the OH group. Thus, 2-bromophenol or o-bromophenol is the correct IUPAC name.

(e) An aromatic ring with a NH₂ group on it is aniline. This NH₂ group defines the 1-position in the ring. Therefore the ethyl group is in the 4-position, or para to the NH₂ group. Thus, 4-ethylaniline or p-ethylaniline is the correct IUPAC name.

(f) The C₆H₅ group that is attached to the cyclohexane ring is a phenyl group; thus the structure is cis-1-chloro-2-phenylcyclohexane.

12.39 Aromatic compounds undergo limited substitution reactions including halogenation, nitration, sulfonation, and alkylation under the proper conditions.

(a) Aromatics do not react with bases; therefore there is no reaction.

(b) Substitution of bromine occurs only in the presence of a metal halide catalyst. This catalyst is not present; so no reaction takes place.

(c) Sulfonation takes place to form:

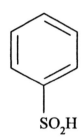

(d) Nitration take place to form:

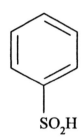

(e) Combustion with O₂ takes place to form:

$$2\ C_6H_6 + 15\ O_2 \rightarrow 12\ CO_2 + 6\ H_2O$$

12.41 The structural formula of benzene is C_6H_6. Therefore, there must be another CHBr on the benzene ring. The possible structures are:

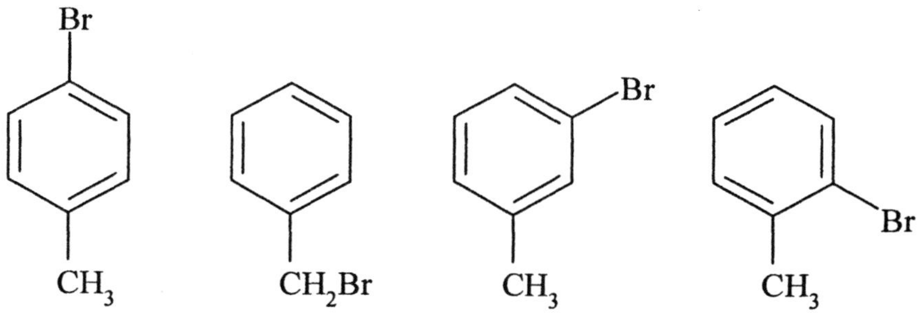

12.43 For a compound to exhibit cis-trans isomers, it must be a cycloalkane or cycloalkene with two single-bonded carbons that each has two different groups attached or an alkene in which each carbon of the double bond has two different substituents attached.

(a) 2-Hexene is an alkene that can form cis-trans isomers:

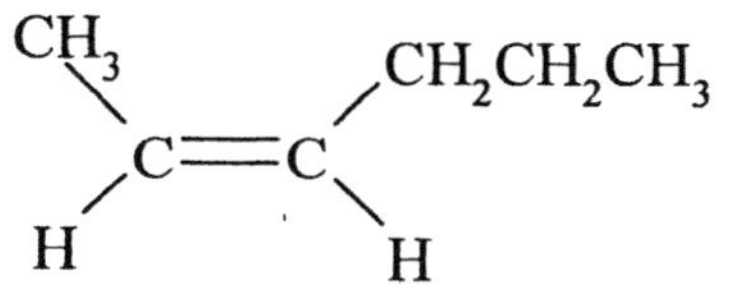

cis

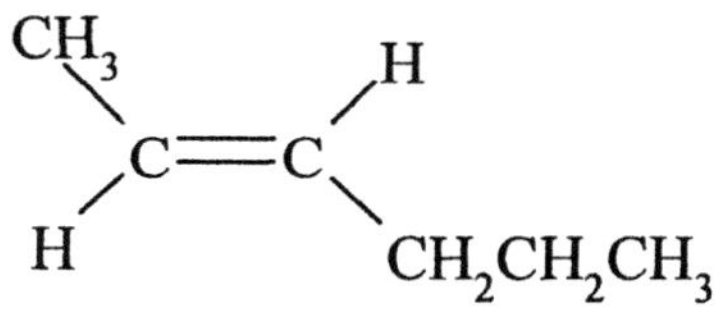

trans

(b) Each of the carbons in a benzene ring has a single substituent, and therefore an aromatic ring cannot exhibit cis-trans isomerism.

(c) In this cyclohexane, two carbons have different substituents to give the isomers:

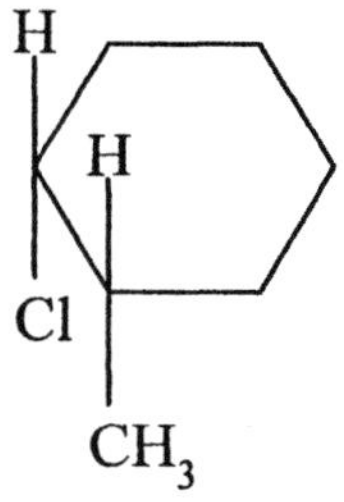

cis

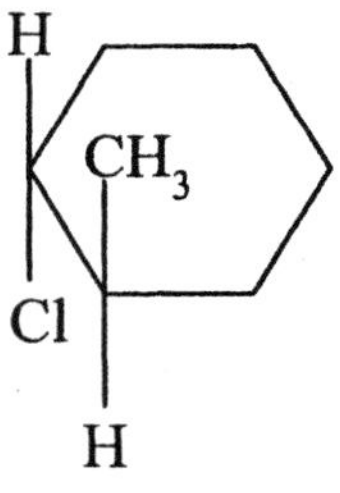

trans

(d) The first and second carbons of a cycloalkene are part of the double bond and can have only one substituent. Therefore, this structure cannot exhibit cis-trans isomerism.

(e) Alkynes have only a single group connected to the carbons that are part of the triple bond (other than the other carbon) and therefore cannot exhibit cis-trans isomerism.

(f) Alkynes have only a single group connected to the carbons that are part of the triple bond (other than the other carbon) and therefore cannot exhibit cis-trans isomerism.

12.45 (a) Both compounds have a formula of C_8H_8; therefore they are isomers or the same compound. The two compounds cannot be superimposed on each other; therefore they are not the same compound. Thus, they are constitutional isomers of each other.

(b) Both compounds have a formula of $C_7H_6Cl_2$; therefore they are isomers or the same compound. The two compounds cannot be superimposed on each other; therefore they are not the same compound. Thus, they are constitutional isomers of each other.

(c) The formula of the first compound is $C_7H_6Cl_2$, whereas that of the second compound is $C_7H_{11}Cl_2$; therefore they must be different compounds that are not isomers.

(d) Both compounds have a formula of $C_7H_6Cl_2$; therefore they are isomers or the same compound. The first compound is aromatic, whereas the other is not; therefore they are not the same compound. Thus, they are constitutional isomers of each other.

(e) Both compounds have a formula of C_4H_8; therefore they are isomers or the same compound. The two compounds cannot be superimposed on each other; therefore they are not the same compound. Thus, they are constitutional isomers of each other.

(f) The formula of the first compound is C_4H_8, whereas that of the second compound is C_4H_6; therefore they must be different compounds that are not isomers.

12.47 All three compounds are nonpolar with very similar boiling points and negligible solubility in water.

12.49 (a) All hydrocarbons will burn (undergo combustion in air) to form CO_2 and H_2O; thus all three will react.

(b) Alkanes, alkenes, and aromatics do not react with a base; therefore none of the three will react.

(c) Only the double bond of cyclohexene will react with an acid such as HCl. The alkane and the aromatic structure are too stable to react under this condition.

(d) Only the double bond of cyclohexene will react with water in the presence of an acid. The alkane and the aromatic structure are too stable to react under these conditions.

(e) Only the double bond of cyclohexene will react with the oxidizing agent $KMnO_4$. The alkane and the aromatic structure are too stable to react under this condition.

12.51 The bromine will react with both double bonds to form:

$$CH_2-CH-CH-CH_2$$
$$\ \ |\ \ \ \ \ \ |\ \ \ \ \ \ |\ \ \ \ \ \ |$$
$$\ \ Br\ \ \ Br\ \ \ Br\ \ \ Br$$

12.53 (a) Counting hydrogens and carbons gives C_6H_8.

(b) Counting hydrogens and carbons gives C_9H_{12}.

(c) Counting hydrogens and carbons gives C_5H_8.

(d) Counting hydrogens and carbons gives $C_8H_7BrClNO_2$.

12.55 An alkyne such as 1-hexyne will undergo an addition reaction to form a dibromoalkene. This alkene can then undergo a further addition reaction to form an alkane, which uses up more bromine. Thus the reaction of the alkyne with Br_2 will utilize twice as much Br_2 as the alkene. This leads to the conclusion that sample A is 1-hexene, whereas sample B is 1-hexyne.

12.57 (a) The double bonds will add the Br_2 to all four carbons to form:

(b) Benzene does not react with Br_2 in the absence of a catalyst; no reaction.

(c) Benzene will undergo a substitution reaction with Br_2 in the presence of $FeBr_2$ to form:

(d) Alkenes do not react with bases; no reaction.

(e) Alkanes do not oxidize in the presence of $KMnO_4$; no reaction.

(f) Benzene will not react with Cl₂ in the presence of heat and light; however, alkanes will. Thus chlorine will replace a hydrogen of the CH₃ group to form:

(g) Benzene will undergo a halogenation reaction in the presence of FeCl₂. Toluene will replace a hydrogen with a chlorine to form the following compounds:

(h) All hydrocarbons will undergo combustion to form CO_2 and H_2O.

(i) All hydrocarbons will undergo combustion to form CO_2 and H_2O.

(j) Benzene is too stable to react with HCl; no reaction.

(k) Cyclohexene will add HCl across the double bond to form:

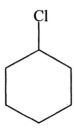

12.59 The change in color signifies a reaction with the dichromate. An alkane is fairly stable and will not react with dichromate, whereas the double bond of an alkene will. Therefore, the unknown must be an alkene, 1-hexene.

Unsaturated Hydrocarbons 81

12.61 Mary Jones is correct. Addition of water to propene will proceed according to Markovnikov's rule to produce 2-propanol as the major product, not 1-propanol.

12.63 Perform the bromine test in a quantitative manner. Weigh out an amount of the shipped compound. Calculate the number of moles assuming the compound is geraniol. Determine the amount of bromine that reacts with the compound. Calculate the number of moles of bromine used. If the compound were geraniol, the number of moles of bromine would be twice the number of moles of the compound. If more than twice the number of moles of bromine are used, the shipped compound is most likely myrcene. This likelihood can be verified by performing the calculation assuming the compound is myrcene. Myrcene would react with three times the number of moles of bromine.

12.65 Unsaturated fats contain carbon–carbon double bonds; saturated fats contain only single bonds.

12.67 The reaction rate increases with increasing stability of the carbocation formed when H^+ adds to the double bond of each alkene. More-stable carbocations are formed faster than less-stable carbocations. 2-Methylpropene forms a tertiary carbocation, propene forms a secondary carbocation, and ethene forms a primary carbocation. The order of carbocation stability $3° > 2° > 1°$ and this results in the observed order of reactivity of alkenes.

Chapter 13

Alcohols, Phenols, Ethers, and Their Sulfur Analogues

13.1 An alcohol is an organic compound that has an –OH bonded to a single nonaromatic carbon. A phenol has an –OH bonded to a carbon that is part of an aromatic ring. An ether is a compound with an oxygen that is bonded to two carbons.

13.3 (a) A tert-butyl group is –C(CH$_3$)$_3$ ≡ C$_4$H$_9$. Therefore, CH$_3$O remains to be bonded to the tert-butyl group. Moreover, because the compound is an alcohol, the CH$_3$O must be bonded as CH$_2$–OH. Therefore, the compound must be:

$$\begin{array}{c} \text{CH}_3 \\ | \\ \text{CH}_3-\text{C}-\text{CH}_2-\text{OH} \\ | \\ \text{CH}_3 \end{array}$$

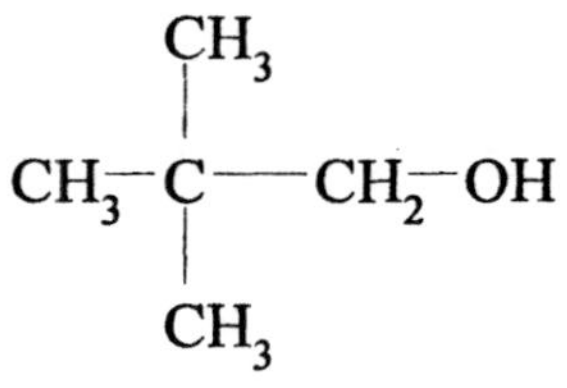

(b) One methyl group = CH$_3$; thus the rest of the compound consists of C$_3$H$_7$O. Because the compound is an alcohol, one of the hydrogens must be bonded to the oxygen. Therefore, the compound must be:

$$\text{CH}_3-\text{CH}_2-\text{CH}_2-\text{CH}_2-\text{OH}$$

13.5 Counting carbons, hydrogens, and oxygens gives:

$$\underset{1}{\text{C}_6\text{H}_{14}\text{O} \equiv \text{C}_n\text{H}_{2n+2}\text{O}} \qquad \underset{2}{\text{C}_6\text{H}_{12}\text{O} \equiv \text{C}_n\text{H}_{2n}\text{O}}$$

13.7 Method for naming alcohols:

(i) Find longest chain (or ring) that contains the –OH. Form base name by adding –ol to the end of the base alkane.

(ii) Put a number before the base name to denote to which carbon the –OH is attached.

(iii) Determine side groups that are attached to the base chain.

(iv) Put names of these groups in alphabetical order at beginning of name. Include numbers to denote to which carbon(s) the groups are attached.

(v) When classifying the alcohol as primary, secondary or tertiary, count the number of carbons that are attached to the carbon on which the –OH is bonded. 1°, 2°, and 3° carbons are directly bonded to one, two, and three other carbon atoms, respectively.

(a) (i) Four carbons = butane → butanol

    (ii) –OH is connected to the first carbon; therefore, 1-butanol.

    (iii) A CH_3 (= methyl) group is connected to C-2.

    (iv) <u>2-Methyl-1-butanol</u>

    (v) That carbon is attached to one other carbon atom; therefore it is a <u>primary alcohol</u>.

(b) (i) Four carbons = butane → butanol

    (ii) –OH is connected to the second carbon; therefore, 2-butanol.

    (iii) A CH_3 (= methyl) group is connected to C-3.

    (iv) <u>3-Methyl-2-butanol</u>

    (v) That carbon is attached to two other carbon atoms; therefore it is a <u>secondary alcohol</u>.

(c) (i) Four carbons = butane → butanol

    (ii) –OH is connected to the first carbon; therefore, 1-butanol.

    (iii) A CH_3 (= methyl) group is connected to C-2.

    (iv) <u>2-Methyl-1-butanol</u>

    (v) That carbon is attached to one other carbon atom; therefore it is a <u>primary alcohol</u>.

(d) (i) Three carbons = propane → propanol

    (ii) –OH is connected to the second carbon; therefore, 2-propanol.

    (iii) A CH_3 (= methyl) group is connected to C-2.

    (iv) <u>2-Methyl-2-propanol</u>

    (v) That carbon is attached to three other carbon atoms; therefore it is a <u>tertiary alcohol</u>.

Chapter 13

13.9 The alcohols have stronger secondary forces than do alkanes and thus have higher boiling points than do alkanes (hydrogen bonds are stronger than London dispersion forces). Therefore, hexane will have the lowest boiling point in this list. Within the group of alcohols, the linear alcohol molecules will interact with other linear alcohols more efficiently than will branched alcohols; thus 2-methyl-2-butanol will have the lowest boiling point of the alcohols. Of the remaining two, 1,4-butanediol will have stronger secondary interactions than does 1-pentanol because it can form two hydrogen bonds per molecule, whereas 1-pentanol can only form one. Thus:

1,4-butanediol B.P. > 1-pentanol B.P. > 2-methyl-2-butanol B.P. > hexane B.P.

13.11 Ethanol is not sufficiently acidic to alter the color of blue litmus paper; therefore there is no change.

13.13 Strong acids will slightly protonate an alcohol. Thus the reaction is:

$$CH_3CH_2CH_2CH_2-OH + HCl \rightleftharpoons CH_3CH_2CH_2CH_2-OH_2^+ + Cl^-$$

13.15 (a) In the absence of heat, alcohols will accept a proton from very strong acids; thus an acid-base reaction will take place:

$$CH_3CH_2CH_2CH_2-OH + H_2SO_4 \rightleftharpoons CH_3CH_2CH_2CH_2-OH_2^+ + HSO_4^-$$

(b) Alcohols do not decompose in the presence of heat; thus no reaction takes place.

(c) In the presence of heat and a strong acid, dehydration, or the loss of water to form a double bond, will take place:

$$CH_2=CH-CH_2-CH_3$$

13.17 A primary alcohol will undergo two consecutive oxidations: first to an aldehyde, and then the aldehyde will be oxidized further to form a carboxylic acid. A secondary alcohol will oxidize to a ketone, whereas a tertiary alcohol will not undergo oxidation short of combustion. Therefore:

(a)

$$CH_3CH_2CH_2CH_2-\overset{O}{\underset{\|}{C}}-H \longrightarrow CH_3CH_2CH_2CH_2-\overset{O}{\underset{\|}{C}}-OH$$

(b) No reaction

13.19 Butene will react with the permanganate at the double bond. 1-Butanol also will react with permanganate to oxidize the –OH. Consequently, both of these compounds will react (and therefore decolor) a permanganate solution. Therefore, this test cannot distinguish between 1-butene and 1-butanol.

13.21 In the naming of phenols, the position of the –OH on the aromatic ring defines the C-1 position on the ring. Additional substituents on the ring should be included in alphabetical order in front of phenol. Thus, this compound is named 2-isopropyl-5-methylphenol.

13.23 When phenol acts as an acid it donates the hydroxyl hydrogen:

$$\text{p-CH}_3\text{-C}_6\text{H}_5\text{—OH} + \text{H}_2\text{O} \rightleftharpoons \text{p-CH}_3\text{-C}_6\text{H}_5\text{—O}^- + \text{H}_3\text{O}^+$$

13.25 (a) Phenols will burn, i.e. undergo combustion in the presence of oxygen:

$$C_6H_6O + 7\,O_2 \rightarrow 6\,CO_2 + 3\,H_2O$$

(b) Phenols are acids and therefore will undergo an acid-base reaction in the presence of a base:

$$C_6H_5\text{–OH} + KOH \rightarrow C_6H_5\text{–O}^- + K^+ + H_2O$$

13.27 (a) 4-Ethoxyphenol: Phenol as the base name means that the primary structure is an aromatic ring with an –OH on it. The carbon to which the –OH is connected is labeled C-1. 4-Ethoxy means that an –O(C$_2$H$_5$) group is connected to the aromatic ring at C-5. Thus:

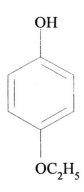

(b) Isobutyl propyl ether: An ether is a compound that connects two organic groups by a single oxygen linkage. The isobutyl and propyl in the name denote that the structures of the two organic groups that exist on both sides of the oxygen; thus:

$$(CH_3)_2CHCH_2—O—CH_2CH_2CH_3$$

(c) 3-Isopropoxy-1-butanol: 1-Butanol means that the base structure is a four-carbon chain (butan-) with an –OH connected to it at C-1. 3-isopropoxy means that an isopropyl group is connected to the main carbon chain by an oxygen (oxy-) linkage at C-3. Thus:

$$\begin{array}{c}OCH(CH_3)_2\\|\\OH—CH_2—CH_2—CH—CH_3\end{array}$$

13.29 The boiling point of a substances is controlled by the strength of its secondary forces; a stronger intermolecular force results in a higher boiling point. Ethanol can form hydrogen bonds between its molecules, whereas the other two compounds cannot; therefore ethanol has the highest boiling point in this list. Between dimethyl ether and propane, dimethyl ether will have dipole–dipole interactions between molecules, whereas propane will have only London dispersion forces; thus:

ethanol B.P. > dimethyl ether B.P. > propane B.P.

13.31 1-Methylcyclopentene is formed at both temperatures. Although at 140°C ether formation is favored for primary alcohols, this alcohol is a tertiary alcohol, which will not undergo a reaction to form an ether.

13.33 (a) 2-Butanethiol: Thiol means that an –SH is connected to the main chain. Butane means that the main carbon chain has four carbons and the 2- denotes that the –SH is connected to the chain at C-2. Thus:

$$\begin{array}{c}CH_3—CH—CH_2—CH_3\\|\\SH\end{array}$$

(b) Dicyclopentyl disulfide: Disulfide means that there exists an –S–S– linkage in this molecule. The dicyclopentyl denotes that the organic structures that are connected to the two ends of the –S–S– linkage are both cyclopentane rings; thus:

Alcohols, Phenols, Ethers and Their Sulfur Analogues

13.35 A, phenol; B, ether; C, alcohol; D, thiol; E, ether.

13.37 An alcohol is a compound that has an –OH bound to an sp^3 carbon, a phenol is a compound that has an –OH group connected to an aromatic hydrogen, a thiol is a compound with an –SH group, an ether is a compound with an oxygen that is bonded to two carbons, and a disulfide is a compound with a –S–S– in it. Thus:

Compound 1 is a disulfide; Compound 2 is a alcohol; Compound 3 is a thiol; Compound 4 is an ether.

13.39 An ether or an alcohol without a double bond or ring must have a formula that fits $C_nH_{2n+2}O$, whereas an alcohol or an ether with a double bond or ring must have a formula that fits $C_nH_{2n}O$. Thus this compound can be either an ether or an alcohol but must have either a double bond or a ring (but not both).

13.41 (a) s-Butyl propyl ether: An ether means that there is an –O– linkage in the main chain with two organic groups on each side. One of them is an s-butyl group, and the other is propyl; thus:

$$CH_3CH_2CH_2-O-CH_2CH(CH_3)_2$$

(b) *m*-Propylphenol: Phenol denotes that the compound is an aromatic with an –OH that defines the C-1 position. *m*-Propyl denotes that there exists a propyl group (C_3H_7) on C-3; thus:

[structure: benzene ring with OH at top and CH₂CH₂CH₃ at meta position]

(c) *cis*-1,2-Cyclohexanediol: Cyclohexanediol denotes that the molecule is a six-carbon ring with two –OH groups attached. The 1,2 denotes that the –OH groups are attached to C-1 and C-2. Cis denotes that that the –OH groups are on the same side of the ring; thus:

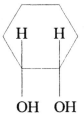

(d) 2-Pentanethiol: Thiol denotes that there is an –SH on the chain, and 2-pentane denotes that the main carbon chain has five carbons and the –SH is connected at C-2; thus:

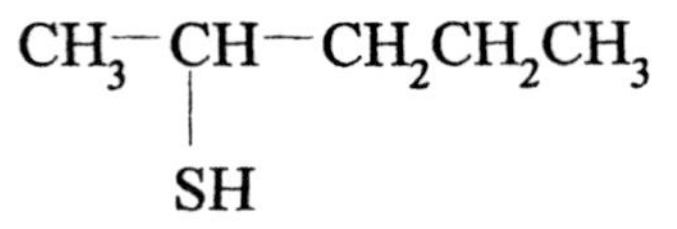

$$CH_3-CH-CH_2CH_2CH_3$$
$$|$$
$$SH$$

(e) 1-Pentene-3-ol: Pentene denotes that the primary structure of the molecule is a five-carbon chain with a double bond. The number 1 denotes that the double bond is between C-1 and C-2. The -3-ol at the end of the name denotes that there exists an –OH on this chain at C-3; thus:

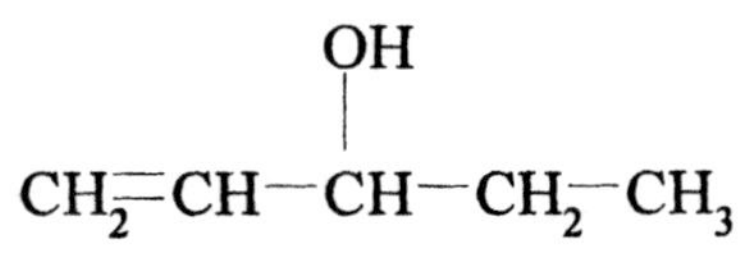

$$OH$$
$$|$$
$$CH_2=CH-CH-CH_2-CH_3$$

13.43 A hydrogen bond will form between an O, N, or F atom and a hydrogen that is covalently bonded to an O, N, or F atom. Thus:

(a) will form hydrogen bonds:

$$C_2H_5-O\cdots H-O$$
with H on first O, C_2H_5 on second O

(b) will form hydrogen bonds:

$$C_3H_7-O\cdots H-O$$
with H on each O, and a second arrangement with C_3H_7-O-H hydrogen bonding to $H-O-H$

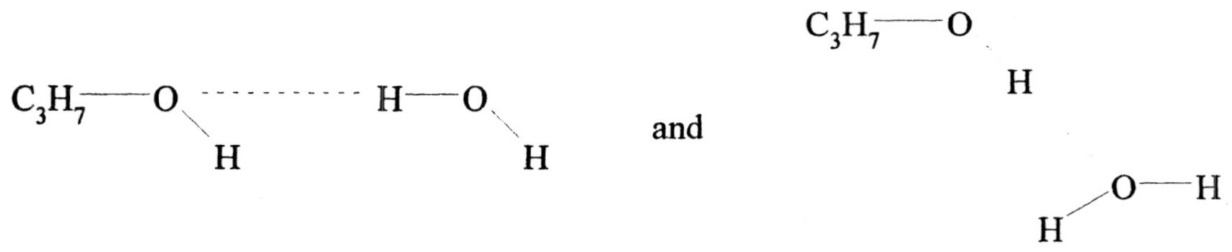

(c) cannot form hydrogen bonds.

(d) will form hydrogen bonds:

$$CH_3-O\cdots H-O$$
with CH_3 on first O, CH_3 on second O

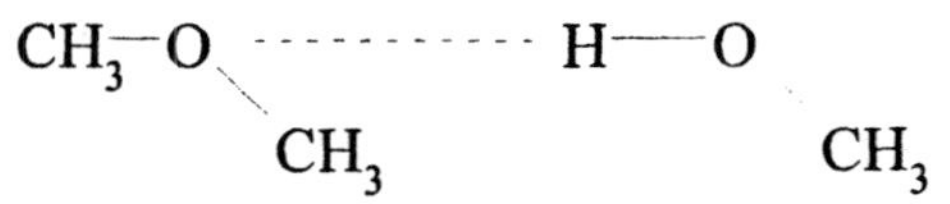

(e) will form hydrogen bonds:

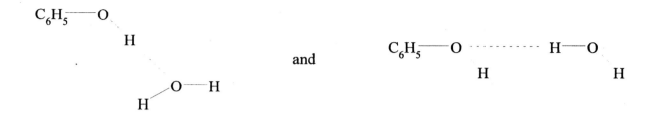

13.45 The molecule that most easily gives up a proton is the most acidic, whereas the one that is least likely to donate a proton is the least acidic. Given this definition, the compounds arranged in decreasing order of acidity are:

HCl > phenol > cyclohexanethiol > cyclohexanol

13.47 (a) Phenols are moderate acids that will react with a very strong base such as NaOH:

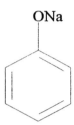

(b) Ethers are neutral and therefore do not react with bases; thus no reaction.

(c) Thiols are weak acids and will thus undergo an acid-base reaction in the presence of a strong base; thus:

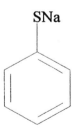

(d) Alcohols are not strong enough acids to react with a base; therefore, no reaction.

(e) Alcohols are very weak acids that will give up a proton to an active metal such as Na:

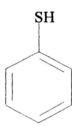 + H$_2$

(f) Thiols will react with heavy metals such as Pb to form a salt:

$$CH_3CH_2CH_2CH_2-S-Pb-S-CH_2CH_2CH_2CH_3$$

(g) The S–H bond in a thiol is weak and therefore a thiol will undergo oxidation to form a disulfide:

$$CH_3CH_2CH_2CH_2-S-S-CH_2CH_2CH_2CH_3$$

(h) Disulfides are easily reduced backed to thiols:

[structure: benzene ring with SH substituent]

13.49 (a) Alcohols are slightly acidic, whereas hydrocarbons are neutral; therefore, 1-propanol.

(b) HCl is a very strong acid, whereas alcohols are merely slightly acidic; therefore, HCl.

(c) Alcohols and water are about the same acidity; therefore the same.

(d) Phenol is more acidic than alcohols owing to the stability of the aromatic ring; thus phenol.

(e) Thiols are more acidic than alcohols; therefore, 1-propanethiol.

13.51 Of the compounds listed here, only 1-butanol will react with permanganate. The fact that the unknown does not react with permanganate means that the unknown is not 1-butanol. The unknown could be any of the other compounds, t-butyl alcohol, ethyl methyl ether, or pentane, because none of these compounds react with permanganate.

13.53   Of these three compound, only phenol is acidic enough to turn blue litmus paper red. Therefore, the unknown must be phenol.

13.55   A, C, and E denote the angle about an oxygen in an ether or an alcohol linkage. This angle is 104.5°. B is the angle about a carbon that is part of an aromatic ring, which equals 120°. Finally, D is the angle about a carbon that is bonded to four atoms, which equals 109.5°.

13.57   Test the compound with blue litmus paper. Hexylresorcinol is a phenol and sufficiently acidic that it turns blue litmus paper red. 3-Hexyl-1, 2-cyclohexanediol is an alcohol and is neutral to litmus paper.

13.59   1-Propanol evaporates rapidly from the skin surface, removing heat and creating a cooling effect. 1-Decanol and 2-decanol are much less volatile because of their higher molecular masses, which results in less heat removed from the skin surface.

13.61   Carbocations (positively charged carbon atoms) are intermediates in the reaction. The carbocation formed from 2-propanol is secondary and more stable than the primary carbocation formed from 1-propanol. The more-stable carbocation is formed faster, which results in a faster overall rate for the dehydration reaction.

13.63   Ethylene oxide is relatively unstable owing to the distortion of its bond angles from the normal tetrahedral bond angle. Reaction to form an acyclic product with tetrahedral bond angles relieves this instability. The six-membered ring of dioxane possesses tetrahedral bond angles and is therefore a very stable compound. Opening of the cyclic dioxane structure does not increase stability.

Chapter 14
Aldehydes and Ketones

14.1 Aldehydes and ketones have the same C:H ratio as alkenes but also have an oxygen, $C_nH_{2n}O$. Alcohols and ethers have molecular formulas of $C_nH_{2n+2}O$. Thus this molecule is an alcohol or an ether.

14.3 An aldehyde has the structure R–CHO, where R denotes a hydrocarbon structure. This molecule has the formula $C_6H_{12}O$; thus the R group must consist of 5 carbons and 11 hydrogens. The isomers that can be formed from this combination are:

$$CH_3CH_2CH_2CH_2CH_2-\overset{O}{\underset{\|}{C}}-H$$

$$CH_3CH_2CH_2-\underset{\underset{CH_3}{|}}{CH}-\overset{O}{\underset{\|}{C}}-H$$

$$CH_3-\underset{\underset{CH_3}{|}}{CH}-CH_2-CH_2-\overset{O}{\underset{\|}{C}}-H$$

$$CH_3CH_2-\underset{\underset{CH_3}{|}}{CH}-CH_2-\overset{O}{\underset{\|}{C}}-H$$

$$CH_3-\underset{\underset{CH_3}{|}}{CH}-CH-\overset{O}{\underset{\|}{C}}-H$$

$$(CH_3)_3CCH_2-\overset{O}{\underset{\|}{C}}-H$$

$$CH_3CH_2-\underset{\underset{CH_3}{|}}{\overset{\overset{CH_3}{|}}{C}}-\overset{O}{\underset{\|}{C}}-H$$

14.5 A ketone has the general formula R–CO–R', where R and R' are hydrocarbon structures, whereas an aldehyde has the structure R–CHO, where R denotes a hydrocarbon structure. Thus compound 1 is an aldehyde. Compound 2 is not an aldehyde or a ketone; it is an ester (from Table 14.1). Compound 3 is a ketone. Compound 4 is not an aldehyde or a ketone; it is an amide (from Table 14.1).

14.7 Atoms a, d, and e are carbons that are bonded to four other atoms; therefore they are sp^3 hybridized. Atoms b and c are atoms that are part of a double bond and are therefore sp^2 hybridized.

14.9 (a) 2,3-Dimethylhexanal: Hexanal means that the base structure is a six-carbon aldehyde. The 2,3-dimethyl denotes that there are two methyl groups connected to this six-carbon chain at C-2 and C-3; thus:

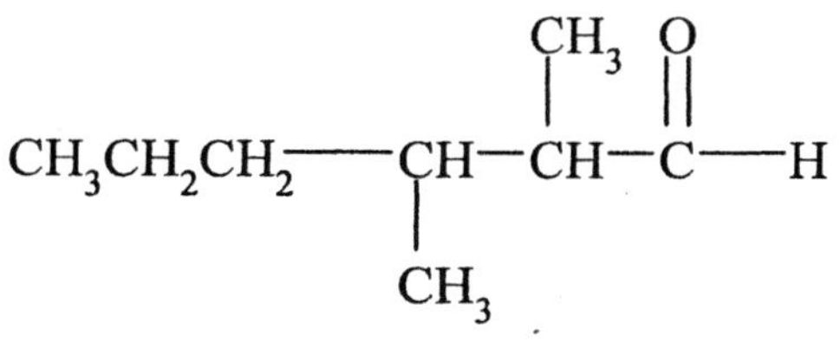

$$CH_3CH_2CH_2-\underset{\underset{CH_3}{|}}{CH}-\underset{\underset{}{|}}{\overset{\overset{CH_3}{|}}{CH}}-\overset{\overset{O}{\|}}{C}-H$$

(b) **Ethyl isopropyl ketone**: Ketone denotes an organic structure that has two hydrocarbon groups on both sides of a carbonyl C=O. The ethyl denotes that one of the groups is an ethyl group, and the isopropyl denotes that an isopropyl group is on the other side of a carbonyl.

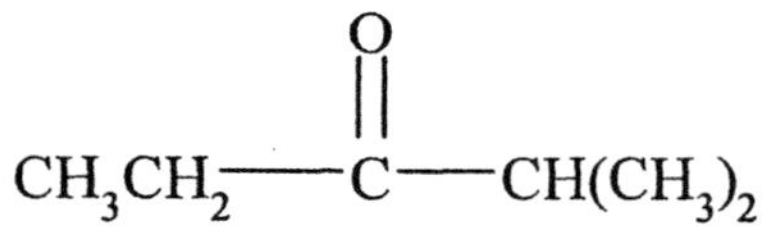

$$CH_3CH_2-\overset{\overset{O}{\|}}{C}-CH(CH_3)_2$$

(c) **3-s-Butyl-4-ethylbenzaldehyde**: Benzaldehyde means that the central structure is a benzene ring with an aldehyde group (CHO) attached to it. The carbon to which the aldehyde group is attached is C-1. 4-Ethyl denotes that an ethyl group is connected to the benzene ring at C-4, and 3-s-butyl denotes that an s-butyl group is connected to C-3.

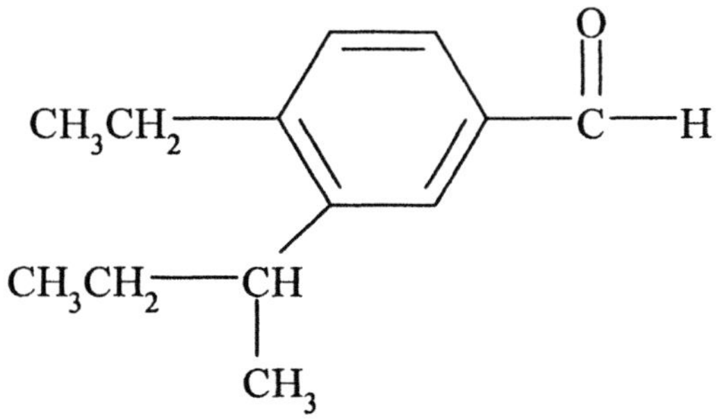

14.11 Method for naming aldehydes:

(i) Find longest chain (or ring) that contains the CH=O. Form the base name by replacing end of alkane name with "al."

(ii) Determine side groups that are attached to this base chain.

(iii) Put names of these groups in alphabetical order at beginning of name. Include numbers to denote to which carbon(s) the groups are attached.

Method for naming ketones:

(0) If the two groups connected to the carbonyl are simple groups, the name is simply the name of the two groups followed by "ketone."

(i) Find longest chain (or ring) that contains the C=O. Form the base name by replacing end of alkane name with "one."

(ii) Count from end to denote which carbon is part of the carbonyl. Be sure to count from end that gives lowest number.

(iii) Determine side groups that are attached to the base chain.

(iv) Put names of these groups in alphabetical order at beginning of name. Include numbers to denote to which carbon(s) the groups are attached.

(a) Use the method for naming aldehydes:

(i) Four carbons → butanal

(ii) An isopropyl group is connected to C-2.

(ii) 2-Isopropylbutanal

(b) Use the method for naming ketones:

(0) The two groups on the sides of the carbonyl are phenyl and *t*-butyl; thus this compound is named *t*-butyl phenyl ketone.

(c) Use the method for naming ketones:

(i) Five-carbon ring → cyclopentanone

(ii) One methyl group is attached to C-2.

(iii) 2-Methylcyclopentanone

(d) Use the method for naming aldehydes:

(i) Five carbons → pentanal

(ii) A chloro group is on C-3 and two methyl groups are on C-4.

(iii) 3-Chloro-4,4-dimethylpentanal

14.13 (a) 3-Pentanone. Both compounds are ketones and therefore have similar secondary forces, but 3-pentanone has a higher molecular mass.

(b) These two compounds will have similar boiling points because aldehydes and ketones of the same molecular mass have about the same intermolecular interactions.

(c) Aldehydes and ketones of the same molecular mass have about the same water solubility because they both can form hydrogen bonds to H_2O at the carbonyl oxygen.

(d) Butanal. Aldehydes have much stronger secondary attractive forces (dipole–dipole) compared with the weaker dispersion forces of hydrocarbons and thus have a higher boiling point.

(e) Butanal. Aldehydes can hydrogen bond with water, but there are no specific attractive forces between pentane and water.

(f) 2-Butanone. Ketones hydrogen bond with water at the carbonyl, whereas the rest of the molecule is essentially a hydrocarbon that does not interact favorably with water. In 2-butanone, the proportion of the molecule that interacts with water is greater than that of 3-pentanone.

(g) 2-Butanol. The alcohol 2-butanol can form strong secondary attractive forces with itself (hydrogen bonds), whereas the ketone 2-butanone can form only the weaker dipole–dipole interactions.

(h) The water solubilities are very nearly the same because both ketones and alcohols can form hydrogen bonds with water.

14.15 Chromate or permanganate will not oxidize compound 1, because ketones are not oxidized under these mild conditions. Compound 2 will be oxidized to a carboxylic acid (CH_3CH_2COOH) because these mild reagents oxidize aldehydes. Finally, compound 3 is a primary alcohol that will be oxidized to an aldehyde, which will be further oxidized to a carboxylic acid.

$$CH_3CH_2CH_2CH_2OH \longrightarrow CH_3CH_2CH_2-\overset{\overset{O}{\|}}{C}-H \longrightarrow CH_3CH_2CH_2-\overset{\overset{O}{\|}}{C}-OH$$

96 Chapter 14

14.17 (a) Tollens's' reagent will oxidize an aldehyde but not a ketone or alcohol. A black or shiny mirror precipitate indicates a positive test with Tollens's' reagent. Compounds 1 and 2 are not aldehydes and therefore will give a negative result with Tollens's' reagent, whereas compounds 3 and 4 are aldehydes and will give a positive result.

(b) Benedict's reagent will oxidize an α-hydroxy ketone or α-hydroxy aldehyde. Compounds 1 and 4 have an –OH connected to the carbon that is next to the carbonyl group in an aldehyde or ketone and therefore will give a positive test with Benedict's reagent. Compounds 2 and 3 will give negative tests. A red precipitate will indicate a positive test.

14.19 Permanganate will oxidize an aldehyde and a primary alcohol and therefore cannot differentiate between 1-pentanol and pentanal.

14.21 A ketone will undergo reduction in the presence of a metal hydride to form a secondary alcohol, whereas an aldehyde will undergo reduction in the presence of a metal hydride to form a primary alcohol; thus:

(a) This compound will undergo reduction to form:

$$CH_3CH_2-\underset{\underset{\displaystyle }{OH}}{CH}-CH_3$$

(b) This compound will undergo reduction to form:

$$CH_3CH_2CH_2OH$$

(c) This alcohol will not undergo reduction.

14.23

$$H^- + H_2O \rightarrow H_2 + {}^-OH$$

H^- is a base in this reaction because it accepts a proton.

14.25 In this reaction, the CH_3OH is first added to the C=O double bond to form the hemiacetal, which then undergoes further substitution reaction to form the acetal:

$$C_6H_5-\overset{O}{\underset{}{C}}-H \xrightarrow{CH_3OH,\ H^+} C_6H_5-\underset{OCH_3}{\overset{OH}{\underset{|}{C}}}-H \xrightarrow[-H_2O]{CH_3OH, H^+} C_6H_5-\underset{OCH_3}{\overset{OCH_3}{\underset{|}{C}}}-H$$

14.27 (a) A linear acetal is formed by the addition and substitution of an alcohol to a carbonyl group of an aldehyde or ketone. To determine the starting materials of a linear acetal, draw the expanded structure of the acetal; then replace the RO–C–OR linkage with a carbonyl group. The alcohol can then be found by adding an H to the RO– group (ROH). Thus:

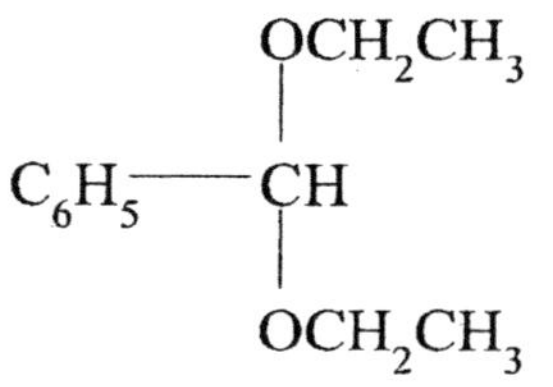

$$C_6H_5-\underset{\underset{OCH_2CH_3}{|}}{\overset{\overset{OCH_2CH_3}{|}}{CH}} \quad \xrightarrow[\text{to get:}]{\text{replace RO-C-OR with } C=O} \quad C_6H_5-\overset{O}{\overset{\|}{C}}H \quad \text{as the starting aldehyde}$$

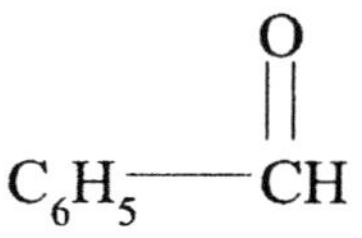

The other starting material, the alcohol, must therefore be CH_3CH_2OH.

(b) In the formation of a cyclic acetal, a cyclic hemiacetal is first formed from a linear compound that has an –OH at one end and an aldehyde carbonyl at the other, as shown in Example 14.10. The hemiacetal then reacts with an alcohol to form the acetal. Thus, for this structure, the hemiacetal that is first formed compound is found by substituting the ether linkage in the acetal that is not in the ring with an –OH:

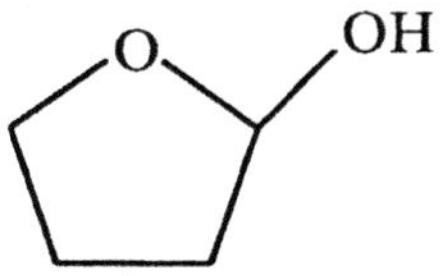

Thus, this hemiacetal must be formed from a four-carbon chain with an aldehyde on one end and an –OH on the other:

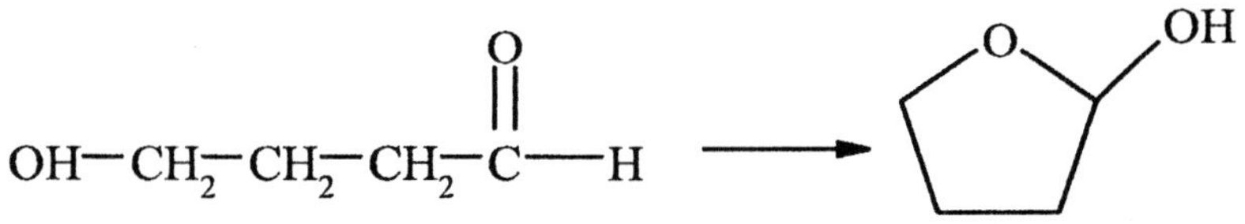

$$OH-CH_2-CH_2-CH_2-\overset{O}{\overset{\|}{C}}-H \longrightarrow$$

This hemiacetal can then react with methanol to form the acetal:

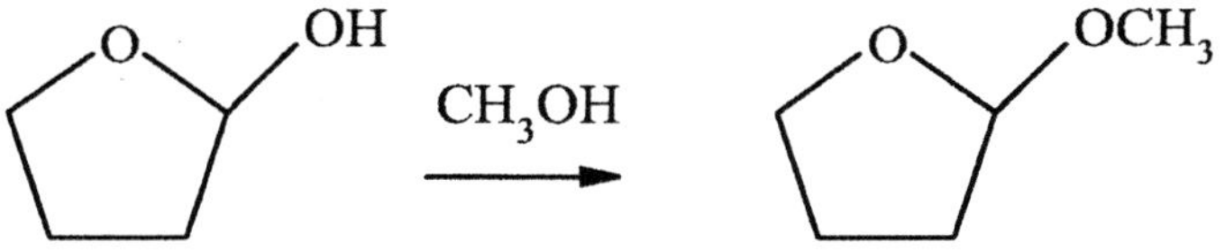

$$\xrightarrow{CH_3OH}$$

(c) This linear hemiacetal is formed by the addition of an alcohol across the carbonyl group, and thus the starting compounds can be found by replacing the HO–C–OR group with a carbonyl to form the starting ketone and recognizing that the ROH must be the starting alcohol:

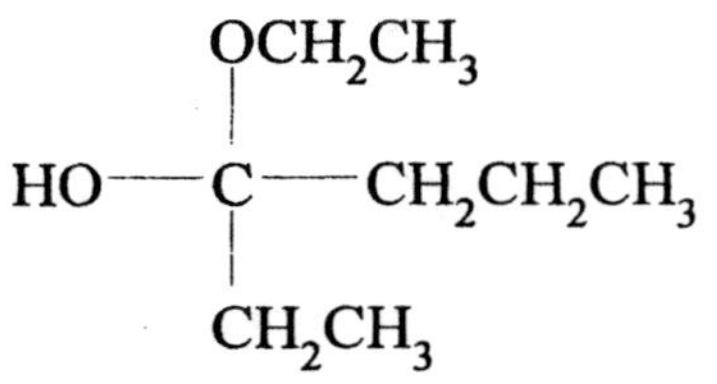

$$HO-\underset{\underset{CH_2CH_3}{|}}{\overset{\overset{OCH_2CH_3}{|}}{C}}-CH_2CH_2CH_3 \quad \xrightarrow{\text{replace the HO-C-OR with C=O to get:}} \quad \underset{\underset{CH_2CH_3}{|}}{\overset{\overset{O}{\|}}{C}}-CH_2CH_2CH_3$$

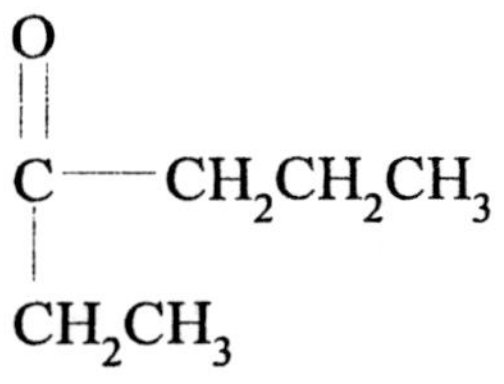

The other starting compound must therefore be the alcohol CH_3CH_2OH.

(d) To complete this part of the exercise, you need to draw the expanded structure of this acetal, which is:

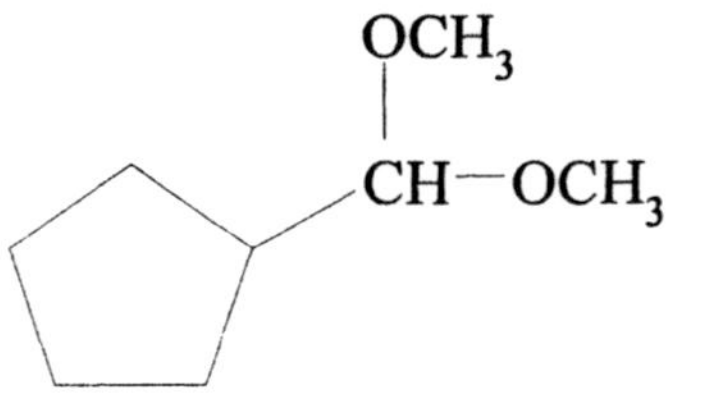

[cyclopentane ring with CH–OCH$_3$ substituent bearing an additional OCH$_3$ group]

Formation of this acetal is by the addition and substitution of an alcohol to a carbonyl group of an aldehyde. To determine the starting materials of this acetal, replace the RO–C–OR linkage with a carbonyl group. The alcohol can then be found by adding an H to the RO– group (ROH). Thus:

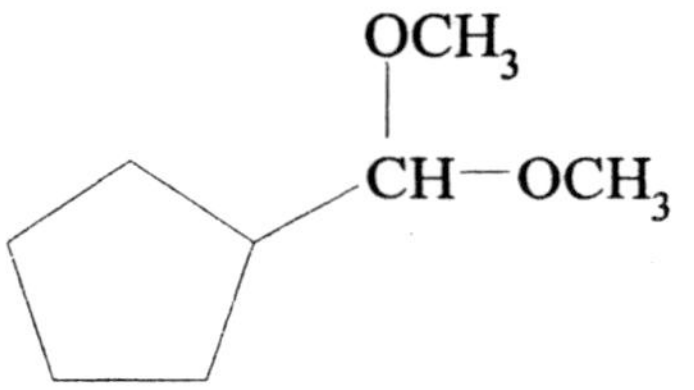

[cyclopentane-CH(OCH$_3$)$_2$]   replace RO-C-OR with C=O to get: [cyclopentane-CHO]

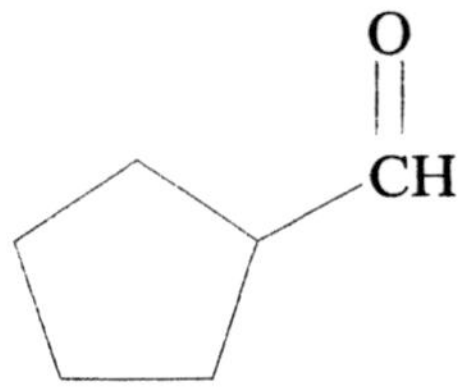

The other starting material, the alcohol, must therefore be CH_3OH.

14.29 (a) The carbon and oxygen atoms that are part of C=C and C=O are sp^2 hybridized.

(b) All bond angles in the C=C and C=O double bonds are 120°.

(c) The C=O double bond is polar because it is a bond between two different atoms, whereas the C=C bond is not.

(d) Both double bonds will undergo a variety of addition reactions— however, not exactly the same reaction under similar conditions.

Aldehydes and Ketones 99

14.31 (a) An aldehyde has the general formula RCHO, where R is a hydrocarbon structure. Therefore, in this molecule, R = C_4H_9. The one methyl group means that this R group has only one end (that is, no branching). Therefore, the structure must be:

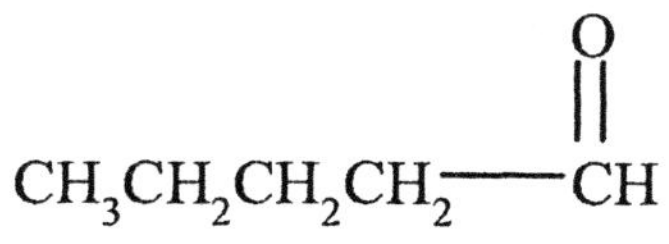

$$CH_3CH_2CH_2CH_2-\underset{\underset{O}{\|}}{C}H$$

(b) This description can be true only if the structure is a five-carbon ring with an =O connected to it; thus:

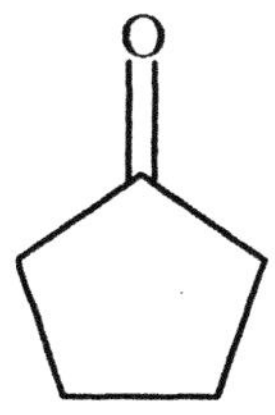

[cyclopentanone structure]

(c) A C_8 ketone denotes that there is a carbonyl group (one carbon and one oxygen) that has two hydrocarbon structures bonded to it. There also exists a benzene ring, which accounts for six of the carbons, leaving a single carbon to make up the hydrocarbon structure on the other side of the carbonyl from the benzene ring; thus:

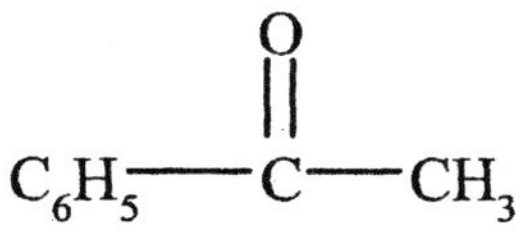

$$C_6H_5-\underset{\underset{O}{\|}}{C}-CH_3$$

(d) $C_6H_{12}O$ → aldehyde = CHO group = one carbon, one oxygen, and one hydrogen. A *t*-butyl group = $C(CH_3)_3$ = four carbons and nine hydrogens, which leaves a single CH_2 group. This CH_2 must be part of the hydrocarbon group and must be between the CHO and *t*-butyl group; thus:

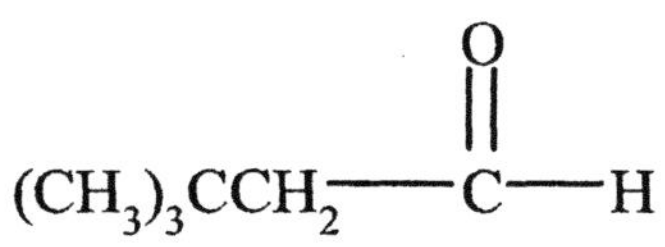

$$(CH_3)_3CCH_2-\underset{\underset{O}{\|}}{C}-H$$

(e) $C_6H_{12}O$ → ketone = a single carbonyl = one carbon and one oxygen. A *t*-butyl group = $C(CH_3)_3$ = four carbons and nine hydrogens, which leaves CH_3 for the other hydrocarbon structure on the other side of the carbonyl from the *t*-butyl group; thus:

100   Chapter 14

$$\underset{CH_3-C-C(CH_3)_3}{\overset{\overset{\displaystyle O}{\|}}{}}$$

14.33 Draw the structure and then rename as in Exercise 14.11

(a) This structure should be named so that the number that denotes where the methyl group is attached to the ring is the lowest possible. Thus it should be 3-methylcyclopentanone.

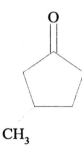

(b) The longest chain in this compound that contains the carbonyl is five carbons long, not four. Thus the name should be 2-ethylpentanal.

$$\underset{\underset{CH_2-CH_2-CH_3}{|}}{\overset{\overset{\displaystyle O}{\|}}{CH-CH-CH_2-CH_3}}$$

(c) The carbonyl cannot be on C-1 in a ketone, which makes it an aldehyde; thus the name of this compound is pentanal.

$$\underset{\underset{O}{\|}}{CH-CH_2-CH_2-CH_2-CH_3}$$

(d) The numbering of the carbons including the carbonyl carbon and the carbon to which the chloro group is attached should be from the other end of the chain to give 3-chloro-2-pentanone.

$$CH_3-CH_2-\underset{\underset{}{\overset{\overset{\displaystyle Cl}{|}}{}}}{CH}-\underset{\underset{}{\overset{\overset{\displaystyle O}{\|}}{}}}{C}-CH_3$$

14.35 (a) 1-Butanol. The alcohol 1-butanol can form strong secondary attractive forces with itself (hydrogen bonds), whereas the aldehyde butanal can form only the weaker dipole–dipole interactions. Thus the boiling point of 1-butanol is higher.

(b) Alcohols and ketones of the same molecular mass have about the same water solubility because they both can form hydrogen bonds with H_2O at the carbonyl oxygen.

(c) Butanal. Aldehydes have stronger dipole–dipole attractive forces from the carbonyl than the dipole–dipole from the C–O–C bond of an ether; thus butanal has a higher boiling point.

(d) The water solubilities are very nearly the same because both aldehydes and ethers can form hydrogen bonds with water.

14.37 The compounds that contain an oxygen can form hydrogen bonds to water and will thus be soluble to a similar extent. Alcohols will hydrogen bond slightly better than aldehydes, which will hydrogen bond slightly better than ethers. The hydrocarbon cannot preferentially interact with water and will therefore be least soluble in water.

$$1\text{-butanol} \approx > \text{butanal} \approx > \text{diethyl ether} > \text{pentane}$$

14.39 Tollens's' reagent will oxidize an aldehyde but not a ketone or alcohol. Therefore, Tollens's' reagent will not distinguish between this ketone and tertiary alcohol.

14.41 The aldehyde in glucose will undergo reduction to form the hexol (a compound with six –OH groups):

$$\text{HOCH}_2-\underset{\text{OH}}{\text{CH}}-\underset{\text{OH}}{\text{CH}}-\underset{\text{OH}}{\text{CH}}-\underset{\text{OH}}{\text{CH}}-\text{CH}_2\text{OH}$$

14.43 The reaction of 2-butanone with excess methanol will first form the hemiacetal by addition of the CH_3OH to the carbonyl double bond. The hemiacetal then undergoes further reaction with methanol to form the more-stable acetal.

$$\underset{CH_3CH_2}{}\overset{O}{\underset{\|}{C}}-CH_3 \xrightarrow{CH_3OH, H^+} CH_3CH_2-\underset{OCH_3}{\overset{OH}{\underset{|}{C}}}-CH_3 \xrightarrow[-H_2O]{CH_3OH, H^+} CH_3CH_2-\underset{OCH_3}{\overset{OCH_3}{\underset{|}{C}}}-CH_3$$

14.45 (a) This ketone will react with excess methanol to form the acetal:

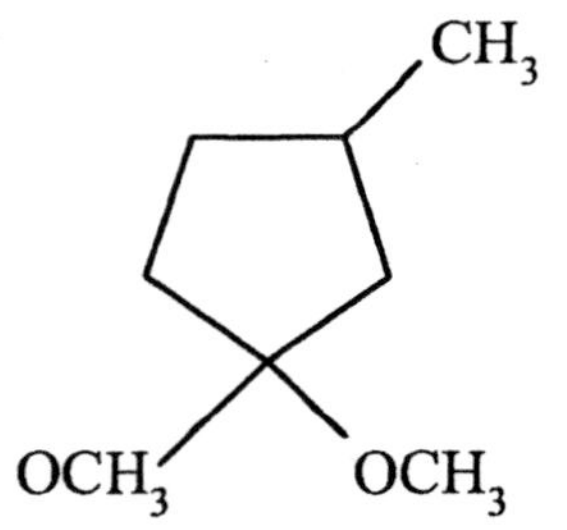

[cyclopentane ring with CH₃ substituent and two OCH₃ groups on the same carbon]

(b) This ketone will undergo reduction in the presence of H_2 and Ni to form:

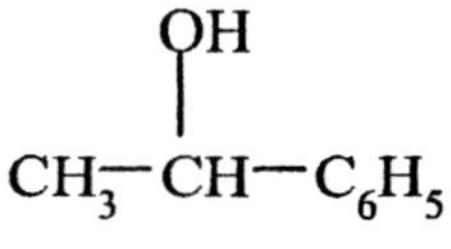

$$CH_3-\underset{\underset{OH}{|}}{CH}-C_6H_5$$

(c) An acetal cannot undergo a reduction reaction, and thus there is no reaction.

(d) Tollens's' reagent will oxidize an aldehyde to a carboxylic acid to form:

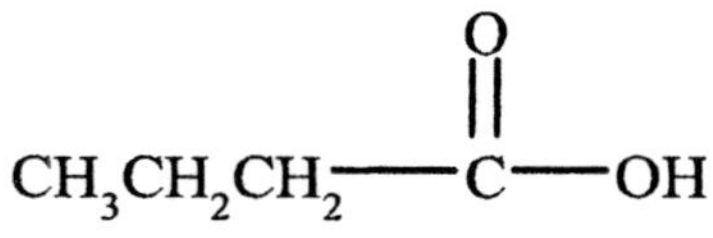

$$CH_3CH_2CH_2-\underset{\underset{}{\overset{\overset{O}{\|}}{C}}}{}-OH$$

(e) An acetal will not undergo an oxidation reaction in the presence of Benedict's reagent, and thus there is no reaction.

(f) NADH acts similarly to a metal hydride to reduce the aldehyde to a primary alcohol:

$$HOCH_2CH_2C_6H_5$$

(g) A ketone will undergo reduction in the presence of a metal hydride and subsequent addition of water to form:

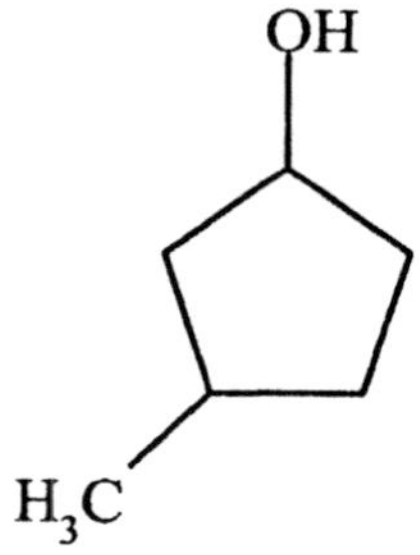

[cyclopentane ring with OH on one carbon and H₃C on another carbon]

(h) An ether will not undergo hydrolysis, and therefore there is no reaction.

14.47 Reduction is equivalent to adding two hydrogens across a double bond; thus:

$$CH_3-\overset{\overset{\displaystyle O}{\|}}{C}-CH_2CH_3$$

14.49 (a) A tertiary alcohol cannot be oxidized to form an aldehyde or ketone.

(b) This primary alcohol will be oxidized to form the following aldehyde:

$$CH_3-\underset{\underset{\displaystyle }{}}{\overset{\overset{\displaystyle CH_3}{|}}{CH}}-CH=O$$

(c) This secondary alcohol will be oxidized to form the following ketone:

(d) Phenol will not undergo oxidation.

14.51 Vanillin is a phenol and is therefore acidic enough to turn blue litmus paper red, whereas the ketone menthone is neutral to litmus paper. Vanillin is also an aldehyde, which shows a positive Tollens's' test, whereas the ketone shows a negative Tollens's' test.

14.53 The hydride reduction reacts only with the carbonyl C=O; however, the catalytic hydrogenation will react with the C=O and the C=C double bonds in cinnamaldehyde, with the loss of two more hydrogens.

14.55 The cyclic compounds are formed by the reaction of the –OH at C-5 with the carbonyl at C-1 to form a stable cyclic hemiacetal. There are two structures because there are two positions (up and down) for the –OH group on C-1 of the hemiacetal.

Chapter 15

Carboxylic Acids, Esters, and Other Acid Derivatives

15.1 A carboxylic acid is a compound with a carbonyl bonded to a hydroxyl group; an ester is a compound with a carbonyl bonded to an oxygen atom; an amide is a compound that has a carbonyl group bonded to an amino group (a nitrogen atom); an anhydride is a compound that has a carbonyl group bonded to two oxygen atoms; and, finally, an acid halide is compound that has a carbonyl group bonded to a halogen (Cl, F, I, and so forth). Thus:

Compound 1 is none of the possibilities listed; it is a ketone and an ether because the two oxygen atoms are not bonded to the same carbon. Compound 2 is an ester. Compound 3 is an acid chloride. Compound 4 is none of the possibilities listed; it is a ketone and an alcohol because the two oxygen atoms are not bonded to the same carbon. Compound 5 is a carboxylic acid. Compound 6 is an amide. Compound 7 is a carboxylic acid and an ether. Compound 8 is an anhydride.

15.3 A carboxylic acid has the structure R–COOH, where R denotes a hydrocarbon structure. This molecule has the formula $C_5H_{10}O_2$; thus the R group must consist of four carbons and nine hydrogens. The isomers that can be formed from these combinations are:

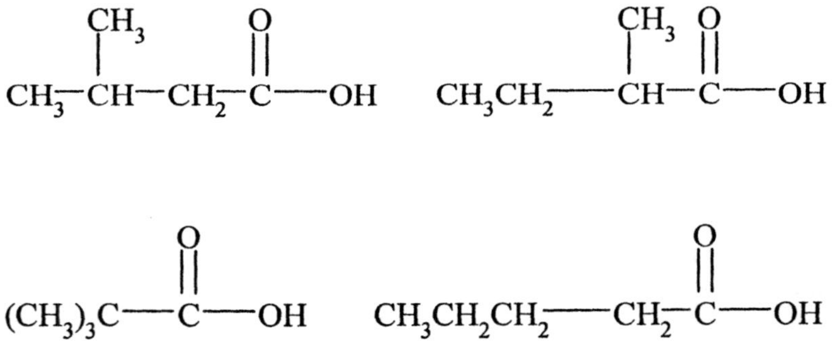

$$CH_3-\underset{\underset{CH_3}{|}}{CH}-CH_2-\overset{\overset{O}{\|}}{C}-OH \qquad CH_3CH_2-\underset{\underset{CH_3}{|}}{CH}-\overset{\overset{O}{\|}}{C}-OH$$

$$(CH_3)_3C-\overset{\overset{O}{\|}}{C}-OH \qquad CH_3CH_2CH_2-CH_2-\overset{\overset{O}{\|}}{C}-OH$$

15.5 Carboxylic acids are synthesized by the selective oxidation of primary alcohols, aldehydes, or an alkyl benzene; thus:

(a) In this alkyl benzene, isopropyl benzene, the alkyl group will be oxidized to a carboxylic acid:

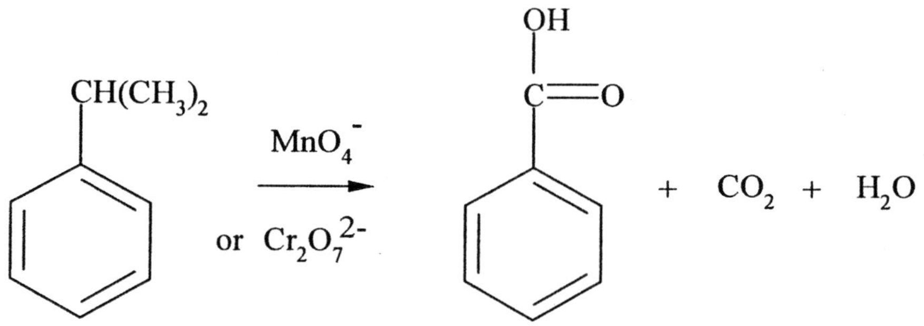

Carboxylic Acids, Esters, and Other Acid Derivatives

[Reaction: isopropylbenzene with MnO_4^- or $Cr_2O_7^{2-}$ → benzoic acid (C(=O)OH on benzene ring) + CO_2 + H_2O]

(b) This primary alcohol will first oxidize to an aldehyde and then to a carboxylic acid:

$$CH_3CH_2CH_2CH_2CH_2\text{—}OH \xrightarrow[\text{or } Cr_2O_7^{2-}]{MnO_4^-} CH_3CH_2CH_2CH_2\text{—}\overset{O}{\overset{\|}{C}}\text{—}OH$$

(c) 2-Pentanol is a secondary alcohol that will oxidize to a ketone, which will not undergo further oxidation.

15.7 Method for naming carboxylic acids:

(i) Find longest chain (or ring) that contains the COOH. Form the base name by replacing end of alkane name with "oic acid."

(ii) Determine side groups that are attached to this base chain.

(iii) Put names of these groups in alphabetical order at beginning of name. Include numbers to denote to which carbon(s) the groups are attached.

(a)

   (i) Longest chain = six carbons = hexanoic acid

   (ii) An isopropyl group is attached to C-2, and two methyl groups are attached to C-2 and C-5.

   (iii) 2-Isopropyl-2,5-dimethylhexanoic acid

(b)

    (i) Longest chain = five carbons = pentanoic acid

    (ii) A chloro group is attached to C-3 and two methyl groups are attached to C-4.

    (iii) 3-Chloro-4,4-dimethylpentanoic acid

(c)

    (i) Longest chain = four carbons = butanoic acid

    (ii) A phenyl group is attached to C-3.

    (iii) 3-Phenylbutanoic acid

(d)

    (i) Longest chain = four carbons = butanoic acid

    (ii) There is another carboxylic acid group (–COOH) on C-2. This is a diacid and is denoted in the name by adding "dioic acid" to the name of the alkane.

    (iii) 1,2-Butanedioic acid

15.9   A pair of carboxylic acid molecules can align to form a dimer that allows two hydrogen bonds between the two molecules:

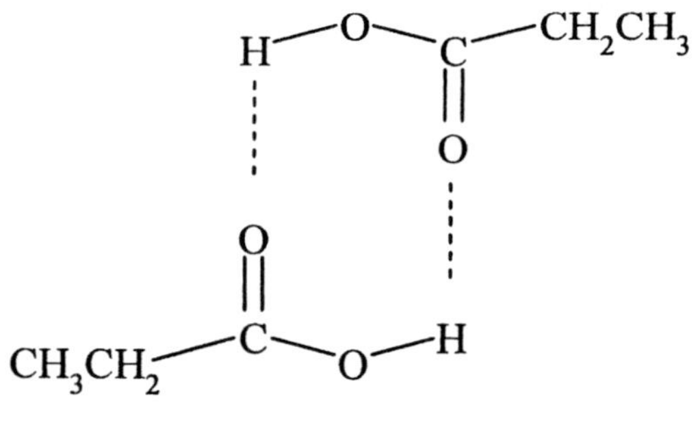

15.11 (a) Two carboxylic acid molecules can align to form a dimer that allows two hydrogen bonds between the two molecules. This "double" hydrogen bonding is a stronger intermolecular interaction than the single hydrogen bond that can form in alcohols; thus ethanoic acid has a higher boiling point than 1-propanol.

    (b) The solubilities of these two compounds are about the same because both can form hydrogen bonds with water.

15.13 When the –OH is directly bonded to the carbonyl, the molecule is a carboxylic acid and more acidic than the alcohol and ketone of compound 2.

15.15 If a compound that is added to a carboxylic acid is a base, an acid base reaction will take place.

(a) Water is a weak base and thus a reaction will take place, but the equilibrium will be far to the left.

$$CH_3-C(=O)-OH + H_2O \rightleftharpoons CH_3-C(=O)-O^- + H_3O^+$$

(b) NaOH is a strong base; thus:

$$CH_3-C(=O)-OH + NaOH \longrightarrow CH_3-C(=O)-O^-Na^+ + H_2O$$

(c) NaHCO$_3$ is a base; thus:

$$CH_3-C(=O)-OH + NaHCO_3 \longrightarrow CH_3-C(=O)-O^-Na^+ + CO_2 + H_2O$$

(d) Water is a weak base and thus a reaction with phenol will take place, but the equilibrium will be far to the left.

$$C_6H_5-OH + H_2O \rightleftharpoons C_6H_5-O^- + H_3O^+$$

(e) NaOH is a strong base that will react with phenol, which is slightly acidic; thus:

$$C_6H_5-OH + NaOH \longrightarrow C_6H_5-O^- + Na^+ + H_2O$$

15.17 Propanoic acid dominates in water because the extent of ionization is less than 2%. To form a solution that has propanoic acid and a pH of 7.0, a substantial amount of buffer must also be added to the solution. The presence of this buffer causes the propanoate ion to be the dominant species.

108 Chapter 15

15.19 Propanoic acid is acidic and will turn blue litmus paper red, whereas the alcohol 1-propanol will not. Thus the unknown is propanoic acid.

15.21 (a) Sodium hexanoate: Hexanoate denotes that the carboxylic acid from which the salt is derived has six carbons. Sodium denotes that the hydrogen of the –COOH is replaced by sodium. The sodium ion has a charge of 1+; thus:

$$CH_3CH_2CH_2CH_2CH_2-\overset{\overset{O}{\|}}{C}-ONa$$

(b) Calcium benzoate: Benzoate denotes that the carboxylic acid from which the salt is derived is benzoic acid. Calcium denotes that the hydrogen of the –COOH is replaced by Calcium. Calcium ion has a 2+ charge; thus:

$$(C_6H_5COO)_2Ca$$

15.23 (a) Carboxylate salts are ionic compounds, whereas carboxylic acids are covalent. The ionic intermolecular interactions of an ionic compound are much stronger than any intermolecular interactions of covalent compounds, even hydrogen bonding. Thus, sodium propanoate has a higher melting point than that of propanoic acid.

(b) Carboxylate salts are ionic compounds that dissociate completely in water. The interaction between these dissociated ions and water is much stronger than the hydrogen bond interaction between a carboxylic acid and water; thus sodium hexanoate is more soluble in water than is octanoic acid.

(c) The polar covalent octanoic acid interacts more strongly with hexane, a nonpolar alkane, than does the ionic carboxylate salt sodium hexanoate.

15.25 If a compound turns red litmus paper blue, the compound is a base. Butanoic acid is an acid, whereas sodium butanoate is a base; thus the compound must be sodium butanoate.

15.27 A soap is a compound that has a molecular structure such that part of the molecule prefers an organic environment, whereas another part of the molecule preferentially interacts with water. Usually one end of the molecule is an ionic compound, whereas the other end is a long hydrocarbon tail. Of the structures that are listed in this problem, only compound 3 fits this description.

15.29 An ester is formed by the reaction of an alcohol and a carboxylic acid in an acidic environment. The overall result is that the acid loses an –OH from the –COOH group, whereas the alcohol loses an –H from the –OH group. The resulting carbonyl and –O groups then link to form the ester. Thus these compounds react as folows:

(a)

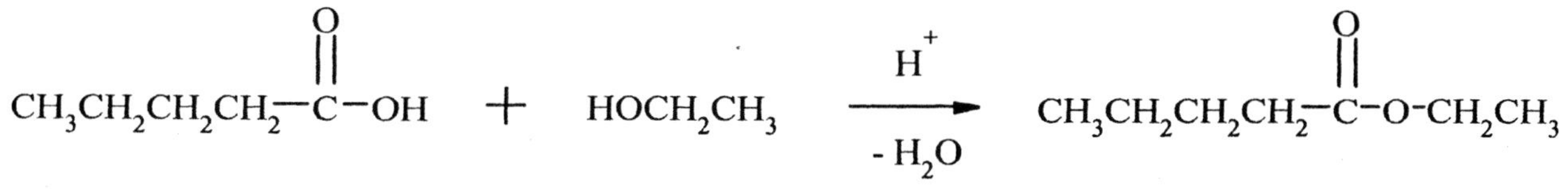

$$CH_3CH_2CH_2CH_2-\overset{\overset{O}{\|}}{C}-OH \ + \ HOCH_2CH_3 \ \xrightarrow[-H_2O]{H^+} \ CH_3CH_2CH_2CH_2-\overset{\overset{O}{\|}}{C}-O-CH_2CH_3$$

(b)

$$C_6H_5-\overset{\overset{O}{\|}}{C}-OH \ + \ HOCH_2CH_2CH_2CH_3 \ \xrightarrow[-H_2O]{H^+} \ C_6H_5-\overset{\overset{O}{\|}}{C}-O-CH_2CH_2CH_2CH_3$$

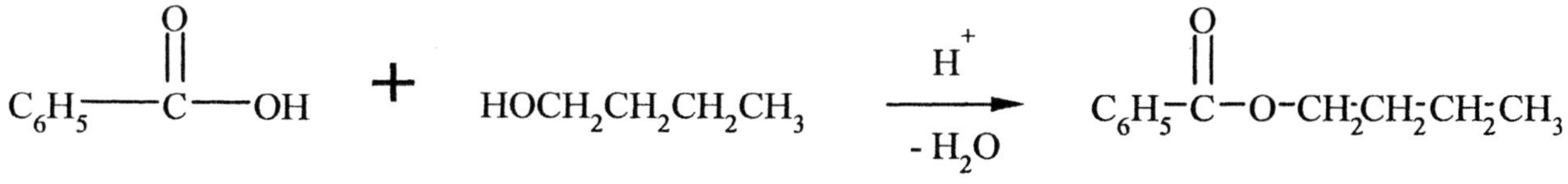

15.31 To determine the starting materials from which an ester is formed, it is best to break the ester linkage between the carbonyl and the oxygen to form two compounds. Then make the starting carboxylic acid from the compound with the –OH group, and make the starting alcohol by adding an H atom to the compound that has the –O end group; thus:

(a) 4,4-Dimethylpentanoic acid and 2-propanol

(b) *p*-Methylbenzoic acid and 2-butanol

(c) Butanoic acid and *p*-methylphenol

15.33 (a) *t*-Butyl butanoate: Butanoate denotes that the carboxylic acid from which the ester is formed is butanoic acid. In other words, the hydrocarbon group that is directly bonded to the carbonyl is a four-carbon straight chain (butane). *t*-Butyl denotes that the alcohol from which the ester is derived is *t*-butyl alcohol. In other words, the hydrocarbon group that is directly bonded to the –O of the ester group is a *t*-butyl group; thus:

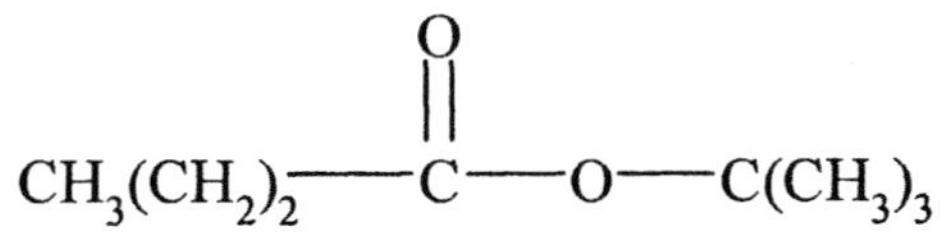

$$CH_3(CH_2)_2-\overset{\overset{O}{\|}}{C}-O-C(CH_3)_3$$

(b) Cyclohexyl benzoate: Benzoate denotes that the carboxylic acid from which the ester is formed is benzoic acid. In other words, the hydrocarbon group that is directly bonded to the carbonyl is a benzene ring. Cyclohexyl denotes that the alcohol from which the ester is derived is cyclohexanol. In other words, the hydrocarbon group that is directly bonded to the –O of the ester group is a cyclohexene ring; thus:

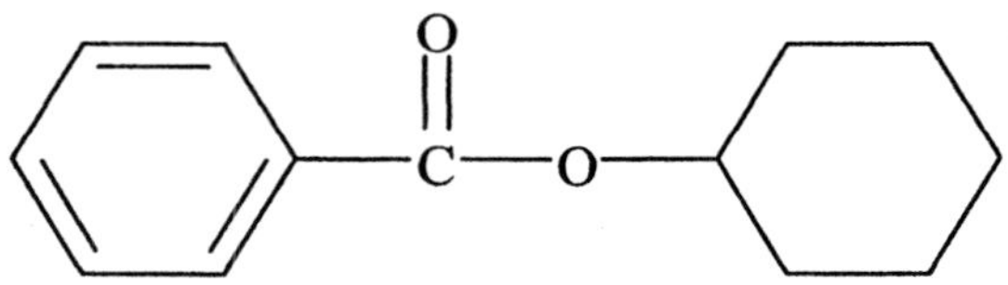

[Structure: phenyl–C(=O)–O–cyclohexyl]

15.35 (a) Both methyl propanoate and pentanal have carbonyl groups that can form dipole–dipole interactions of similar strength. However, the carbonyl group in the ester methyl propanoate, is in the middle of the molecule, whereas the carbonyl group of the aldehyde pentanal is at the end of the molecule. Therefore it will be easier for two dipoles of the aldehyde to come together to form an intermolecular interaction than for the same thing to happen between two ester molecules. Therefore, pentanal has more secondary interactions and thus a higher boiling point.

(b) Water can just as easily hydrogen bond with the carbonyl in an ester as with the carbonyl of an aldehyde; therefore these two molecules will have similar solubilities in water.

(c) The hydrogen bonding in 1-pentanol is a stronger secondary force than the dipole–dipole interactions of the ester methyl propanoate. Thus 1-pentanol has the higher boiling point.

(d) Water can hydrogen bond with the carbonyl or –OH group in a carboxylic acid, whereas it can hydrogen bond only to the carbonyl group of an ester. Therefore, hexanoic acid is more soluble in water than methyl pentanoate.

(e) The hydrogen bonding in butanoic acid is a stronger secondary force than the dipole–dipole interactions of the ester methyl propanoate. Thus butanoic acid has the higher boiling point.

15.37 Polyesters can be formed by the reaction of diacids with diols, similar to the way in which esters are formed from acids and alcohols. Therefore, the diol is HO–R–OH and the diacid has the structure HOOC–R'–COOH. The exact structure of R and R' from a polyester structure can be found by noticing that R is the hydrocarbon group that is between the two oxygen atoms that are part of the carbonyl, whereas R' is the hydrocarbon structure that is between the carbonyls. Thus, in this compound, R = $CH_2CH_2CH_2$ and R' = $CH_2CH_2CH_2CH_2$. Therefore, this polyester is synthesized from 1,6-hexanedioic acid and 1,3-propanediol.

15.39 The ester linkage can be broken under acidic conditions to form the original acid and alcohol or under basic conditions to form a carboxylate salt and an alcohol (or phenol). Thus, these compounds undergo the following reactions:

(a)

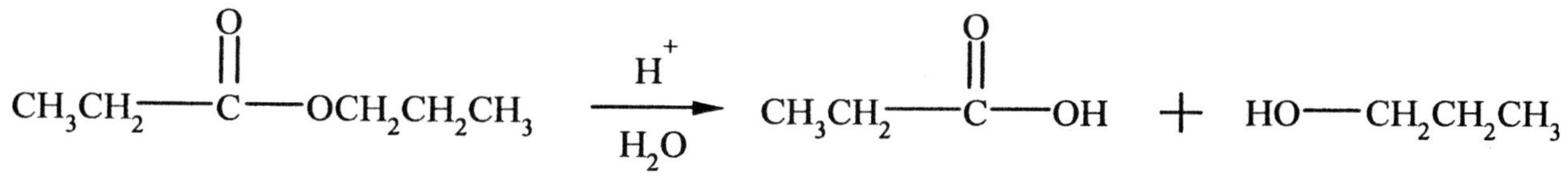

$$CH_3CH_2-\overset{\overset{O}{\|}}{C}-OCH_2CH_2CH_3 \xrightarrow[H_2O]{H^+} CH_3CH_2-\overset{\overset{O}{\|}}{C}-OH \ + \ HO-CH_2CH_2CH_3$$

(b)

$$CH_3CH_2CH_2-\overset{\overset{O}{\|}}{C}-O-C_6H_5 \xrightarrow[H_2O]{NaOH} CH_3CH_2CH_2-\overset{\overset{O}{\|}}{C}-ONa \ + \ HO-C_6H_5$$

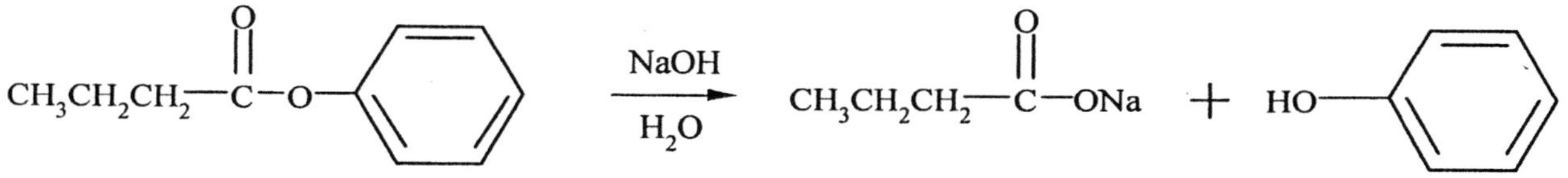

15.41 (a) Acid halides are compounds in which the –OH of a carboxylic acid is replaced by a halogen and are named by replacing the "ic acid" with "yl halide." Thus, for this compound, the –OH of 2-methylpropanoic acid is replaced by a chlorine to form:

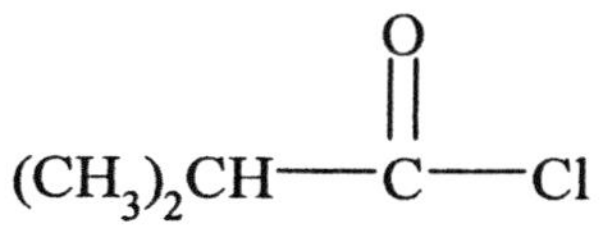

$$(CH_3)_2CH-\overset{\overset{O}{\|}}{C}-Cl$$

(b) An anhydride is a compound that has two carbonyl groups that are bonded to a single oxygen molecule as R–(C=O)–O–(C=O)–R. The R structure is denoted by the name of the anhydride; that is, ethanoic denotes that the R groups are CH_3 in this structure; thus:

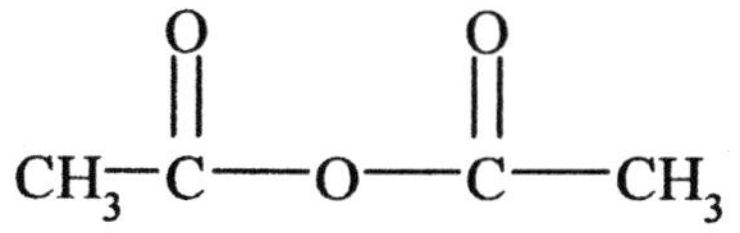

$$CH_3-\overset{\overset{O}{\|}}{C}-O-\overset{\overset{O}{\|}}{C}-CH_3$$

15.43 Acid chlorides react with water to form the corresponding carboxylic acid:

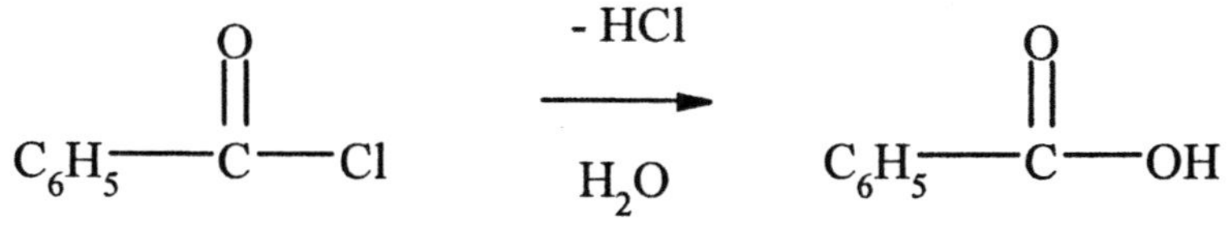

$$C_6H_5-\overset{\overset{O}{\|}}{C}-Cl \xrightarrow[H_2O]{-HCl} C_6H_5-\overset{\overset{O}{\|}}{C}-OH$$

15.45 The HO–P=O bond in phosphoric acid behaves similarly to a carboxylic acid group by undergoing esterification reaction with alcohols. Thus a phosphoric ester will be formed by the reaction of an alcohol and a phosphoric acid. The overall result is that the acid loses an –OH from the –POOH group, whereas the alcohol loses an –H from the –OH group. The resulting –P=O and –O groups then link to form the ester. Thus:

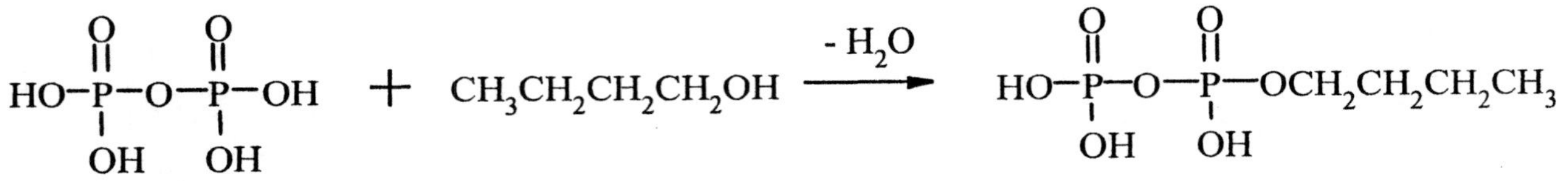

$$HO-\underset{\underset{OH}{|}}{\overset{\overset{O}{\|}}{P}}-O-\underset{\underset{OH}{|}}{\overset{\overset{O}{\|}}{P}}-OH \;+\; CH_3CH_2CH_2CH_2OH \;\xrightarrow{-H_2O}\; HO-\underset{\underset{OH}{|}}{\overset{\overset{O}{\|}}{P}}-O-\underset{\underset{OH}{|}}{\overset{\overset{O}{\|}}{P}}-OCH_2CH_2CH_2CH_3$$

This compound is butyl diphosphate

15.47 Diphosphoric acid gives up its hydrogens that are part of –OH groups in the presence of a base. The only way that a solution of an acid can have a pH of 7 is if there is also a base present to buffer the action of the acid; therefore:

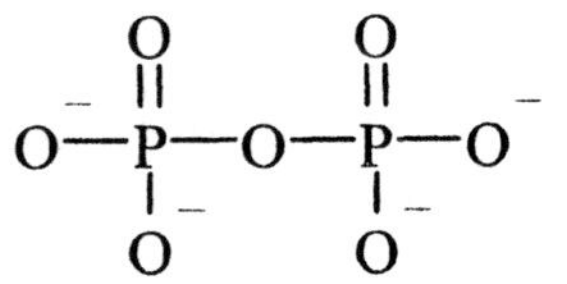

$$\overset{-}{O}-\underset{\underset{O^-}{|}}{\overset{\overset{O}{\|}}{P}}-O-\underset{\underset{O^-}{|}}{\overset{\overset{O}{\|}}{P}}-O^-$$

15.49 *cis*-2-Methylcyclopentane carboxylic acid: 2-Methylcyclopentane denotes a five-membered ring with a methyl group attached to the ring at C-2. The carboxylic acid denotes that a –COOH group is attached to the ring at C-1; thus:

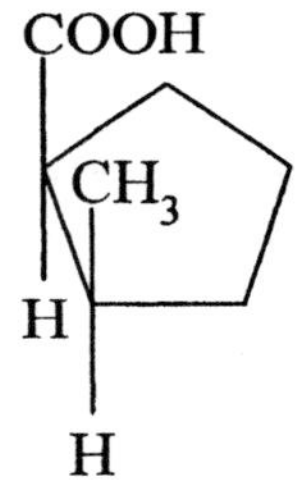

[Structure: cyclopentane ring with COOH and CH₃ on adjacent carbons, both H's shown on the same face indicating cis configuration]

15.51 5-Hydroxypentanoic acid is:

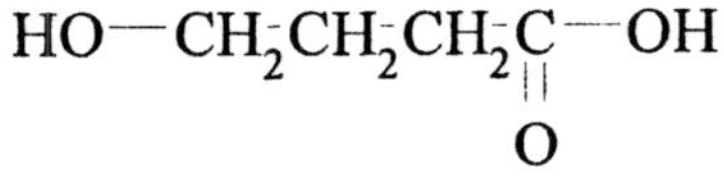

$$HO-CH_2CH_2CH_2CH_2\underset{\underset{O}{\|}}{C}-OH$$

Thus, if the –OH group on one end wraps around to react with the acid group on the other end to form an ester linkage, it will form a six-membered ring (including the oxygen atom):

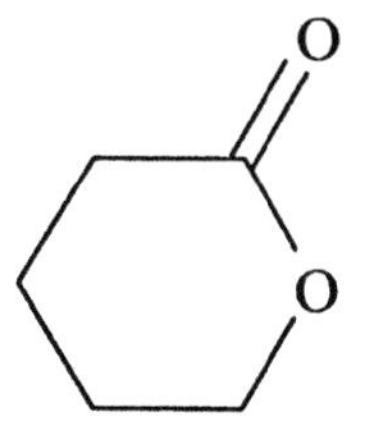

[Structure: six-membered ring lactone (δ-valerolactone) with ring O and C=O]

15.53 (a) *trans*-3-Chlorocyclohexanecarboxylic acid: Cyclohexane denotes that the base structure is a six-membered ring. Carboxylic acid at the end denotes that there is a –COOH group attached to the ring at C-1. *trans*-3-Chloro denotes that there is a chloro group on C-3 on the opposite side of the ring from the COOH group:

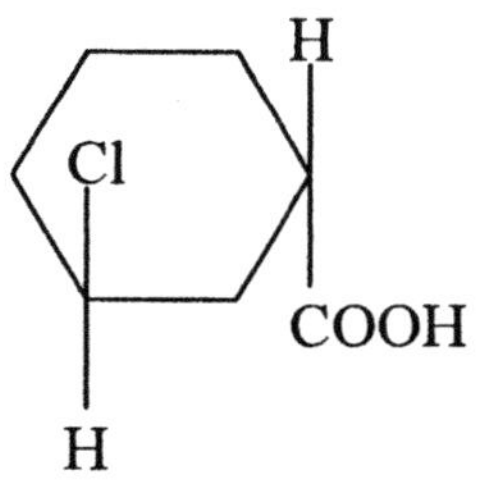

[Cyclohexane ring with Cl and H on C-3 (Cl up, H down) and COOH and H on C-1 (COOH down, H up)]

(b) Calcium acetate: Acetate denotes the carboxylate salt of acetic acid, CH₃COOH. Calcium is the counterion. Because calcium has a 2+ charge and the acetate ion has a 1– charge, there must be two acetates for each calcium ion.

$$(CH_3COO)_2Ca$$

(c) Isopropyl benzoate: This compound is an ester with R–COO–R'. R' is the hydrocarbon structure denoted by the first group in the name, isopropyl, whereas R is the base structure of the carboxylic acid from which the ester can be formed. This is denoted in the name as the base name of the second group in the name benzoate → benzoic acid; thus:

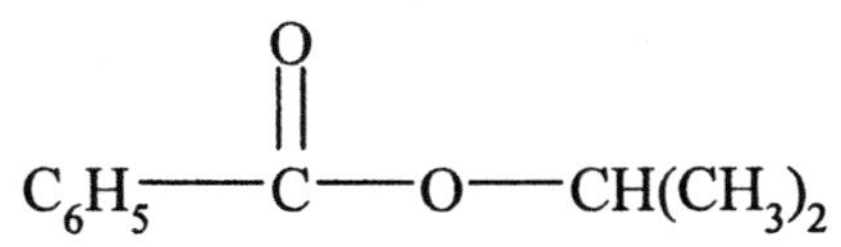

$$C_6H_5-\underset{\underset{\displaystyle}{\|}}{\overset{\overset{\displaystyle O}{\|}}{C}}-O-CH(CH_3)_2$$

(d) Pentanoic anhydride: An anhydride is a compound that has two carbonyl groups that are bonded to a single oxygen molecule as R–(C=O)–O–(C=O)–R. The R structure is denoted by the name of the anhydride; that is pentanoic denotes that the R groups are CH₃CH₂CH₂CH₂ in this structure; thus:

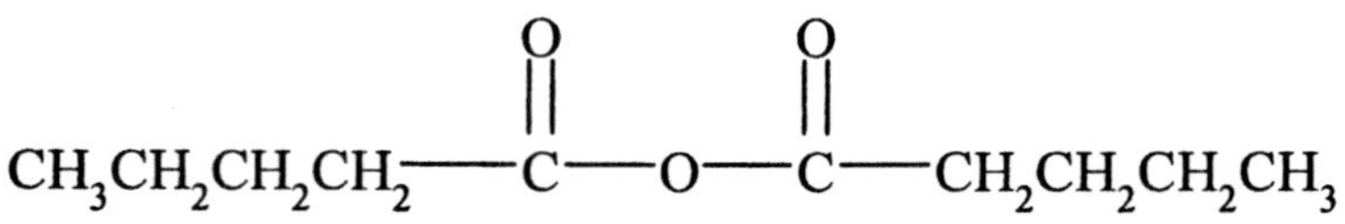

$$\text{CH}_3\text{CH}_2\text{CH}_2\text{CH}_2 - \overset{\overset{\text{O}}{\|}}{\text{C}} - \text{O} - \overset{\overset{\text{O}}{\|}}{\text{C}} - \text{CH}_2\text{CH}_2\text{CH}_2\text{CH}_3$$

(e) Propanoyl chloride: The "yl chloride" denotes an acid chloride in which the –OH of a carboxylic acid is replaced by a –Cl. Propanoyl denotes that the the base hydrocarbon has three carbons; thus:

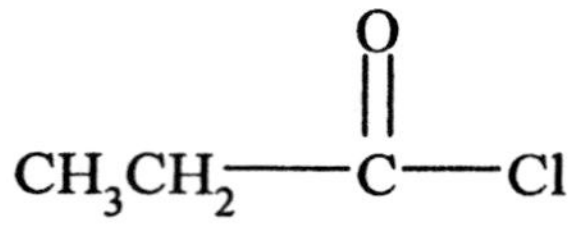

$$\text{CH}_3\text{CH}_2 - \overset{\overset{\text{O}}{\|}}{\text{C}} - \text{Cl}$$

(f) Methyldiphosphate: Diphosphate denotes a phosphoric acid ester with two phosphorous atoms. Methyl denotes that one of the hydrogens of an –OH groups is replaced by a methyl group; thus:

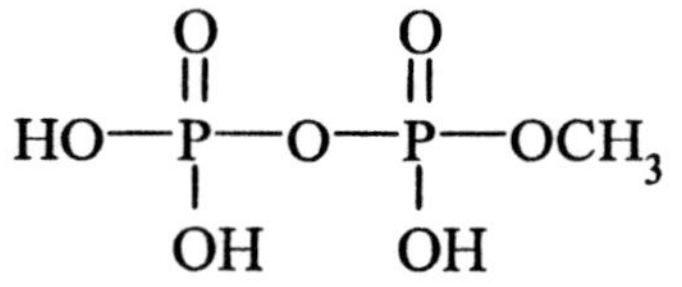

$$\text{HO}-\underset{\underset{\text{OH}}{|}}{\overset{\overset{\text{O}}{\|}}{\text{P}}}-\text{O}-\underset{\underset{\text{OH}}{|}}{\overset{\overset{\text{O}}{\|}}{\text{P}}}-\text{OCH}_3$$

(g) Dimethyl phosphate: Phosphate denotes a phosphoric acid ester with one phosphorous atom. The dimethyl denotes that two of the hydrogens are replaced with CH₃ groups:

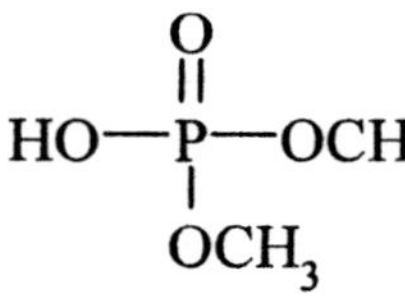

$$\text{HO}-\underset{\underset{\text{OCH}_3}{|}}{\overset{\overset{\text{O}}{\|}}{\text{P}}}-\text{OCH}_3$$

15.55 Atoms a and c are sp^3 hybridized, whereas atoms b and d are sp^2 hybridized. The angles A and C are about the carbon of a carbonyl and are thus 120°, whereas angle B is between C–O–C bonds, which is 104.5°.

15.57 If a compound turns blue litmus paper red, the compound is an acid. Of these three molecules, both phenol and propanoic acid are strong enough acids to turn blue litmus paper red, whereas propanol is not. Thus the unknown is either phenol or propanoic acid.

15.59 An ester is formed by the reaction of a thiol and a carboxylic acid in an acidic environment. The overall result is that the acid loses an –OH from the –COOH group and the alcohol loses an –H from the –SH group. The resulting carbonyl and –S groups then link to form the ester. Thus, these compounds react as follows:

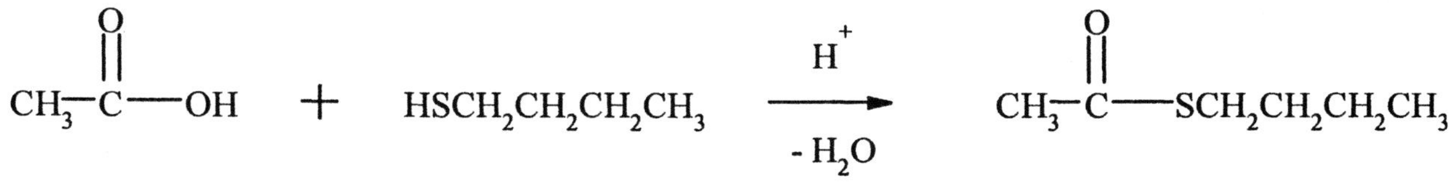

$$CH_3-\overset{O}{\overset{\|}{C}}-OH + HSCH_2CH_2CH_2CH_3 \xrightarrow[-H_2O]{H^+} CH_3-\overset{O}{\overset{\|}{C}}-SCH_2CH_2CH_2CH_3$$

15.61 (a) An ester will not react with water if there is no acid present to catalyze the reaction; thus there is no reaction under these conditions.

(b) An ester hydrolyzes in the presence of water and acid to from an alcohol and carboxylicacid:

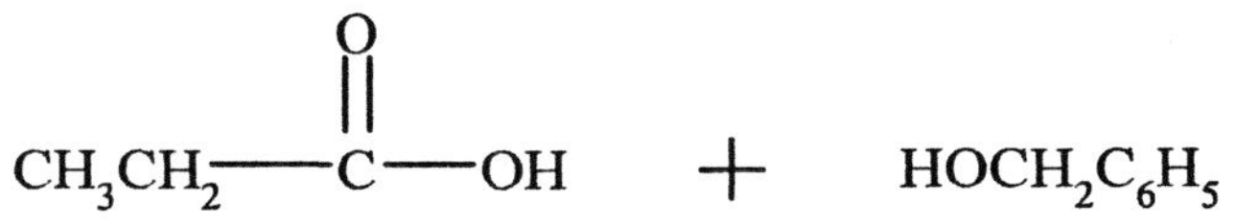

$$CH_3CH_2-\overset{O}{\overset{\|}{C}}-OH + HOCH_2C_6H_5$$

(c) An ester forms an alcohol and carboxylate salt in the presence of water and a base:

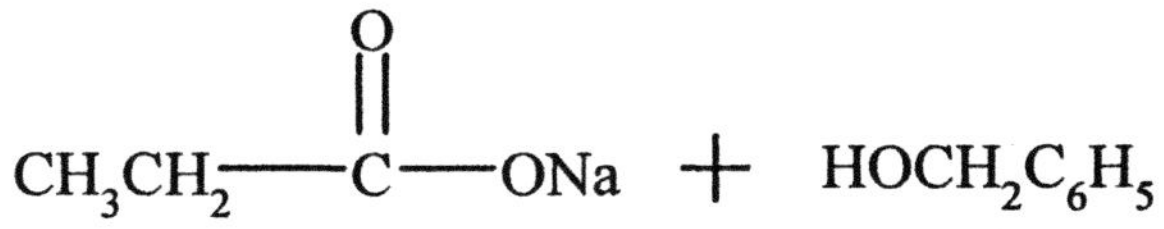

$$CH_3CH_2-\overset{O}{\overset{\|}{C}}-ONa + HOCH_2C_6H_5$$

(d) A lactone is an ester that undergoes a reaction to form an alcohol and a carboxylate salt. However, because the initial compound is a ring, the alcohol and carboxylate salt are on opposite ends of the same molecule:

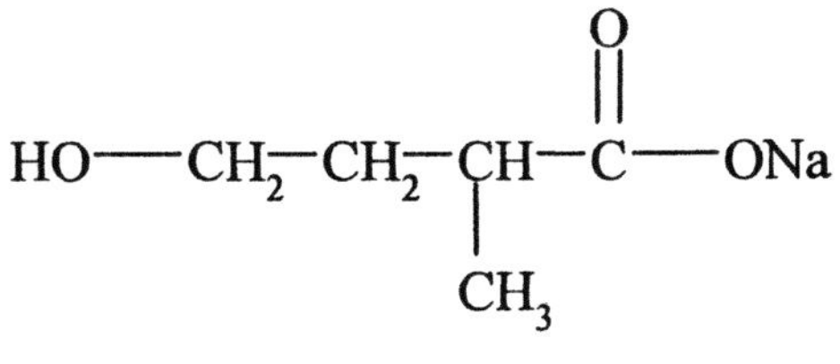

$$HO-CH_2-CH_2-\underset{\underset{CH_3}{|}}{CH}-\overset{O}{\overset{\|}{C}}-ONa$$

(e) A phosphoric acid reacts with an alcohol to form the phosphoric ester:

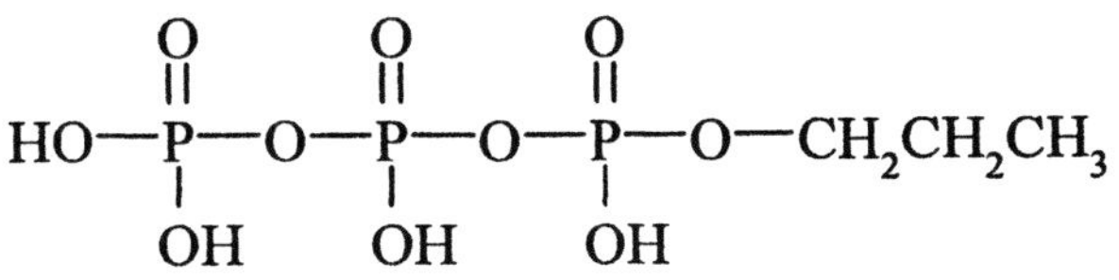

$$HO-\underset{\underset{OH}{|}}{\overset{\overset{O}{\|}}{P}}-O-\underset{\underset{OH}{|}}{\overset{\overset{O}{\|}}{P}}-O-\underset{\underset{OH}{|}}{\overset{\overset{O}{\|}}{P}}-O-CH_2CH_2CH_3$$

(f) A carboxylic acid undergoes an acid-base reaction in the presence of a base to form a carboxylate salt:

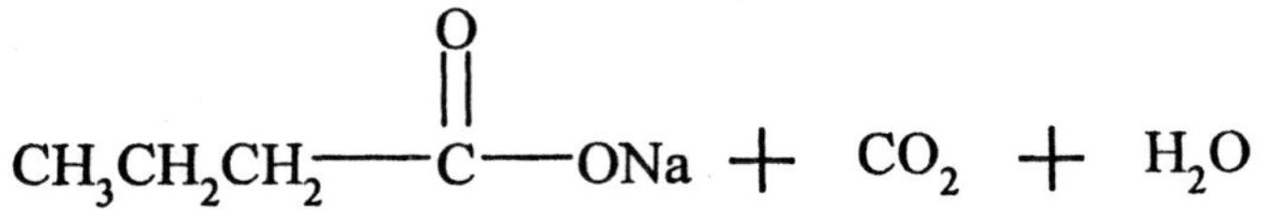

$$CH_3CH_2CH_2-\overset{\overset{O}{\|}}{C}-ONa \;+\; CO_2 \;+\; H_2O$$

(g) An alcohol and a carboxylic acid react under acidic conditions to form an ester:

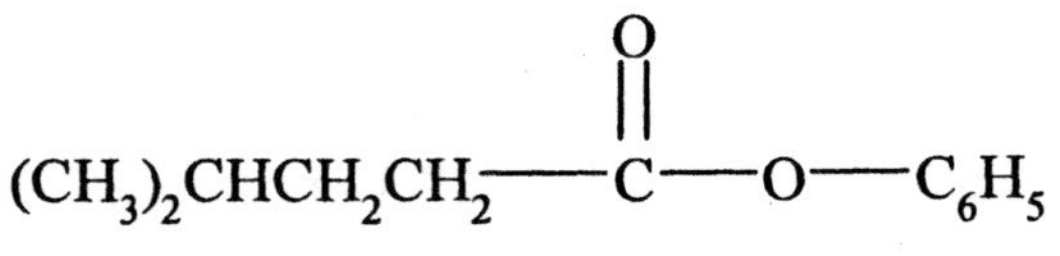

$$(CH_3)_2CHCH_2CH_2-\overset{\overset{O}{\|}}{C}-O-C_6H_5$$

(h) An anhydride reacts with water to form a carboxylic acid:

$$CH_3CH_2-\overset{\overset{O}{\|}}{C}-OH$$

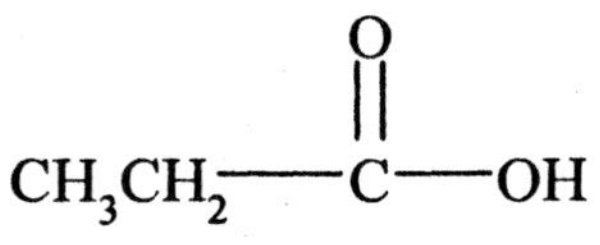

(i) An acid chloride reacts with water to form a carboxylic acid:

$$CH_3CH_2-\overset{\overset{O}{\|}}{C}-OH$$

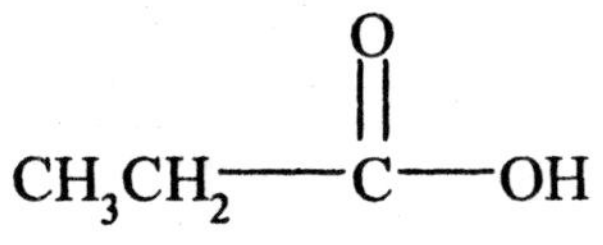

(j) A thioester undergoes a reaction in the presence of water and acidic conditions to form a thiol and a carboxylic acid:

$$CH_3CH_2-\overset{\overset{O}{\|}}{C}-OH \;+\; HSCH_2C_6H_5$$

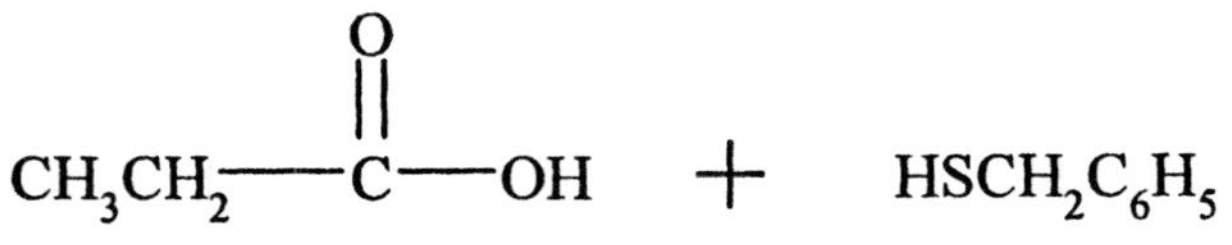

15.63 A carboxylic acid is acidic, whereas an ester is not. Therefore if a drop of the unknown is placed on blue litmus paper and if it is an ester, it will not turn the paper red. However, if the litmus paper turns red, then the unknown is a carboxylic acid.

15.65

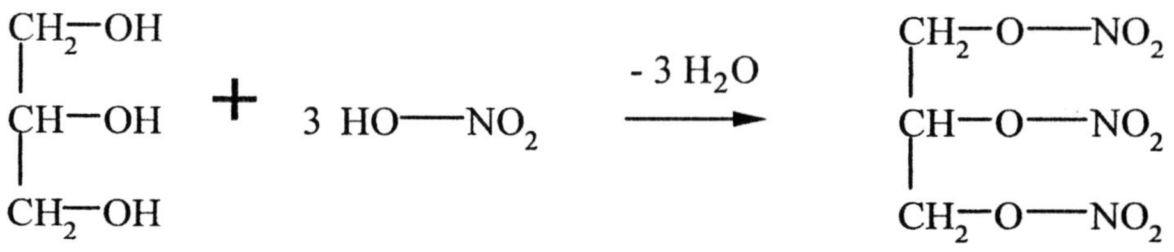

$$\begin{array}{c}CH_2-OH\\|\\CH-OH\\|\\CH_2-OH\end{array} \quad + \quad 3\ HO-NO_2 \quad \xrightarrow{-3\ H_2O} \quad \begin{array}{c}CH_2-O-NO_2\\|\\CH-O-NO_2\\|\\CH_2-O-NO_2\end{array}$$

15.67 As in Box 9.1, if $[H_3O]^+ = x$, $x^2 = 1.74 \times 10^{-5}$; $x = 4.17 \times 10^{-3}$. $pH = -\log[H_3O] = 2.38$.

Chapter 16

Amines and Amides

16.1 A tertiary amine means that the nitrogen is bonded to three carbons. Thus, the isomers of $C_5H_{13}N$ must be defined by where the other two carbons that are part of this compound are bonded to the NC_3 center. There are three possible combinations:

$$CH_3CH_2-\underset{\underset{CH_3}{|}}{N}-CH_2CH_3 \qquad CH_3CH_2CH_2-\underset{\underset{CH_3}{|}}{N}-CH_3 \qquad (CH_3)_2CH-\underset{\underset{CH_3}{|}}{N}-CH_3$$

16.3 An amide is a compound with a nitrogen that is bonded to a carbonyl group, whereas an amine is a compound with a nitrogen that is bonded to carbons or hydrogens but not a carbonyl. In a primary amine the nitrogen is bonded to one carbon and two hydrogens; in a secondary amine, the nitrogen is bonded to two carbons and one hydrogen; and, finally, in a tertiary amine, the nitrogen is bonded to three carbons.

Compound 1 is an amine. Because the nitrogen is bonded to two carbons, it is a secondary amine. Because the nitrogen is part of a nonaromatic carbon chain, it is an aliphatic amine.

Compound 2 is an amine. Because the nitrogen is bonded to one carbon, it is a primary amine. Because the nitrogen is part of a nonaromatic carbon ring, it is an aliphatic amine. There is also a ketone group in this compound.

Compound 3 is an amine. Because the nitrogen has three bonds to carbons, it is a tertiary amine. Because the nitrogen is part of an aromatic carbon ring, it is an aromatic amine.

Compound 4 is an amide.

Compound 5 is an amide because the nitrogen is bonded to a carbonyl.

Compound 6 is neither an amide nor an amine, because the nitrogen is bonded to an oxygen as well as a benzene ring; thus it is a nitrobenzene.

16.5 (a) N-Methyl-1-pentaneamine: 1-Pentaneamine denotes that a nitrogen is bonded to C-1 of a five carbon (pentane) chain. N-Methyl denotes that a methyl group also is bonded to the nitrogen of the amine; thus:

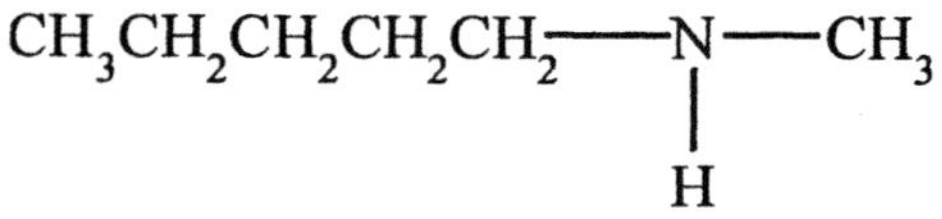

CH3CH2CH2CH2CH2—N—CH3
H

(b) Isopropylmethylamine: Methylamine denotes that a methyl group is bonded to the nitrogen of an amine, and the isopropyl in the name denotes that an isopropyl group also is bonded to the nitrogen; thus:

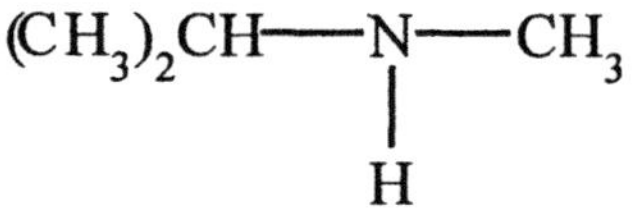

(CH3)2CH—N—CH3
H

(c) *cis*-4-Ethyl-N-propylcyclohexaneamine: Cyclohexanamine denotes that a cyclohexane ring is bonded to the nitrogen of an amine. N-Propyl signifies that a propyl group also is bonded to the nitrogen of the amine. *cis*-4-Ethyl denotes that an ethyl group is bonded to C-4 of the cyclohexane ring (C-1 is the carbon bonded to the N) and it is situated on the same side of the ring as the amine; thus:

H
H
CH2CH3
NHCH2CH2CH3

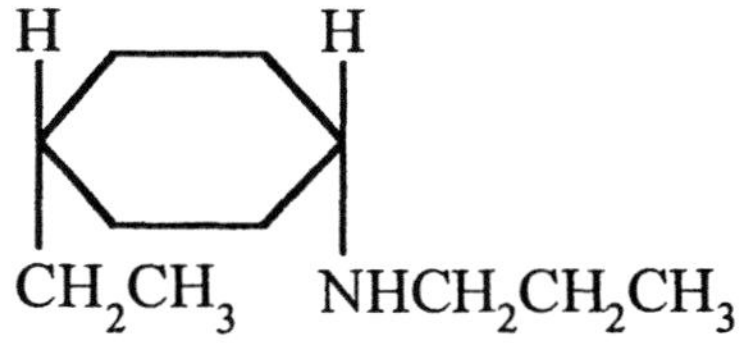

(d) N-Ethylaniline: Aniline is benzene with an NH_2 group connected to it. The N-ethyl denotes that an ethyl group is bonded to the nitrogen of the amine; thus:

HN—CH2CH3

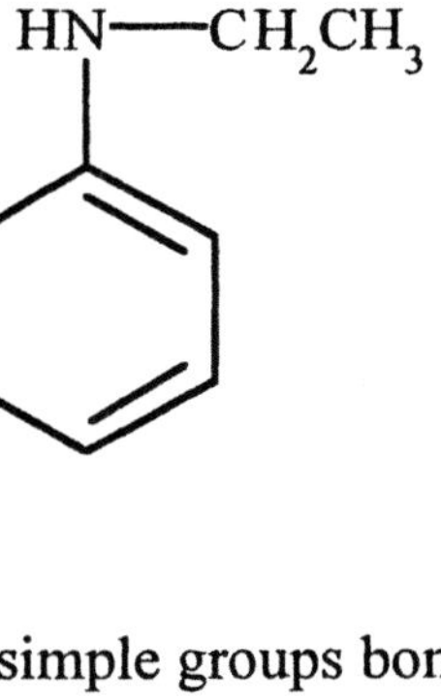

16.7 (a) This is a tertiary amine with three simple groups bonded to the nitrogen. These groups are two methyl groups and an ethyl group. Putting the group names in front of the word "amine" completes naming the amine; thus this compound is ethyldimethylamine.

(b) This compound has a cyclohexane ring bonded to the amine group; thus the base name is cyclohexanamine. There is also a methyl group bonded to the cyclohexane ring at C-2, trans to the amine; thus this compound is *trans*-2-methylcyclohexanamine.

(c) The groups bonded to the nitrogen are not simply named; thus the method used in part a will not work for this compound. The longest chain that has a carbon to which the nitrogen is bonded is a five-carbon chain, pentane. On the pentane chain is a methyl group at C-4 and the amine at C-2. Thus, the base name is 4-methyl-2-pentanamine. The compound is slightly more complex because it has a methyl and an ethyl group on the nitrogen, which is denoted in the name by putting N-methyl and N-ethyl in front of the base name. Because there are two methyl groups, they are combined to form the name N-ethyl-N,4-dimethyl-2-pentanamine.

(d) The nitrogen bonded to the benzene ring is aniline. The ethyl and methyl groups that are bonded to the nitrogen are denoted by putting N-ethyl and N-methyl in front of aniline; thus this compound is N-ethyl-N-methylaniline.

16.9 An amine can form a hydrogen bond between one of the amino hydrogens and the nitrogen of another molecule:

$$C_2H_5-N(H)(H) \cdots H-N(H)-C_2H_5$$

16.11 (a) The O–H bond in an alcohol is more polar than the N–H bond of an amine. This increased polarity results in stronger hydrogen bonds between two alcohol molecules than the hydrogen bonds that exist between two amine molecules. This stronger intermolecular interaction results in 1-butanol having a higher boiling point than butylamine.

(b) The hydrogen bond between water and the N–H of an amine is similar to the hydrogen bond between water and the carbonyl oxygen of an aldehyde. This results in the solubilities of butylamine and butanal being about the same.

(c) The primary amine butylamine can form hydrogen bonds between two molecules because a hydrogen is bonded to the nitrogen. The tertiary amine ethyldimethylamine, however, cannot form hydrogen bonds. Therefore, the intermolecular forces between butylamine molecules are greater than those of ethyldimethylamine, which results in butylamine having a higher boiling point than ethyldimethylamine.

(d) Both primary and secondary amines can hydrogen bond with water; thus they have similar solubilities in water.

16.13 A base can be defined as a proton acceptor. Therefore, propylamine must take the hydrogen from HCl to form a salt:

$$CH_3CH_2CH_2NH_2 + HCl \rightarrow CH_3CH_2CH_2NH_3^+ \; Cl^-$$

16.15 Basicity measures the ability of a compound to abstract a proton. Amines are more basic than water or an alcohol because the nitrogen of the amine has two nonbonded electrons that can bond to a proton. Aniline is less basic than propylamine because the electron-withdrawing effect of the aromatic ring makes the nonbonded electrons of the nitrogen less available. KOH is the strongest base because the base is actually OH$^-$, which seeks the positive charge of the proton to neutralize the anion. Thus,

$$\text{water} \approx \text{propanol} < \text{aniline} < \text{propylamine} < \text{KOH}$$

16.17 A compound that turns red litmus paper blue is a base; thus the unknown must be propylamine.

16.19 A base can be defined as a proton acceptor and each NH$_2$ group will act as a base toward HCl. Therefore, 1,6-hexanediamine must take the hydrogen from HCl to from a salt as:

$$H_2NCH_2CH_2CH_2CH_2CH_2CH_2NH_2 + 2 \; HCl \rightarrow Cl^- \; ^+H_3NCH_2CH_2CH_2CH_2CH_2CH_2NH_3^+ \; Cl^-$$

16.21 (a) N,N-Dimethylcyclopentanammonium chloride: Ammonium chloride denotes an ammonium salt, and the counterion is chloride. Cyclopentanammonium signifies that the base structure of the amine from which the salt was derived had an amine bonded to a cyclopentane ring. Finally, N,N-dimethyl denotes that two methyl groups are bonded to the nitrogen. Thus the structure is:

(b) N-Ethyl-N-methylpiperidinium sulfate: Piperidinium sulfate denotes that this is an ammonium salt that is derived from piperidine (a six-membered ring with a nitrogen in the ring) and the counterion is chloride. The N-ethyl and N-methyl denote that a methyl group and an ethyl group are bonded to the nitrogen. Thus the compound is:

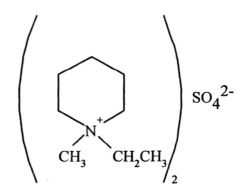

(c) Ethylmethylpropylammonium bromide: Ammonium chloride denotes that this is an ammonium salt, and the counterion is bromide. The ethylmethylpropyl before the ammonium in the name denotes that the amine from which this salt was derived had an ethyl, a methyl, and a propyl group bonded to the nitrogen; thus the compound is:

$$\text{CH}_3\text{CH}_2-\overset{\overset{\displaystyle\text{CH}_3}{|}}{\underset{\underset{\displaystyle\text{CH}_2\text{CH}_2\text{CH}_3}{|}}{\overset{+}{\text{N}}}}-\text{H} \quad \text{Br}^-$$

16.23 (a) An amine salt is an ionic compound. Ionic interactions are much stronger than the hydrogen-bonding secondary forces of an amine; thus trimethylammonium chloride has a higher boiling point than that of triethylamine.

(b) Amine salts can completely dissociate into ions, which interact more strongly with water than an amine can by hydrogen bonding. Thus, hexylammonium chloride has a higher solubility in water than does hexylamine.

16.25 An amine is a base that will turn red litmus paper blue, whereas an ammonium salt is an acid that will turn blue litmus paper red; thus this compound is an amine, butylamine.

16.27 An amide is a compound in which a nitrogen is bonded to a carbonyl group. Amides are classified as primary when two hydrogens are bonded to the nitrogen, as secondary when one hydrogen is bonded to the nitrogen, and as tertiary when no hydrogens are bonded to the nitrogen.

Compound 1 has a nitrogen bonded to a carbonyl; thus it is an amide. There are no hydrogens on the nitrogen; thus it is a tertiary amide.

Amines and Amides   123

Compound 2 does not have a nitrogen bonded to a carbonyl. Thus it is not an amide; it is a ketone and an amine.

Compound 3 has a nitrogen bonded to a carbonyl; thus it is an amide. There are two hydrogens on the nitrogen; thus it is a primary amide. The NH$_2$ group at the other end of the molecule is a primary amine.

Compound 4 has a nitrogen bonded to a carbonyl; thus it is an amide. There is one hydrogen on the nitrogen; thus it is a secondary amide.

16.29 (a) A carboxylic acid will react with an amine at room temperature to undergo an acid-base reaction to form an ammonium salt:

$$C_6H_5-\overset{O}{\underset{\|}{C}}-OH \; + \; H_2NCH_2CH_2CH_2CH_3 \; \xrightarrow{25°C} \; C_6H_5-\overset{O}{\underset{\|}{C}}-O^- \; H_3\overset{+}{N}CH_2CH_2CH_2CH_3$$

(b) A carboxylic acid and an amine will undergo a coupling reaction to form an amide at elevated temperature (greater than 100°C):

$$C_6H_5-\overset{O}{\underset{\|}{C}}-OH \; + \; H_2NCH_2CH_2CH_2CH_3 \; \xrightarrow[-H_2O]{>100\,°C} \; C_6H_5-\overset{O}{\underset{\|}{C}}-NH-CH_2CH_2CH_2CH_3$$

(c) An acid chloride is more reactive than a carboxylic acid; therefore an acid chloride will react with an amine at room temperature to form an amide:

$$C_6H_5-\overset{O}{\underset{\|}{C}}-Cl \; + \; H_2NCH_2CH_3 \; \xrightarrow{-HCl} \; C_6H_5-\overset{O}{\underset{\|}{C}}-NH-CH_2CH_3$$

(d) An anhydride will react with an amine to form an amide and a carboxylic acid:

$$H_3C-\overset{O}{\underset{\|}{C}}-O-\overset{O}{\underset{\|}{C}}-CH_3 \; + \; H\overset{CH_3}{\underset{|}{N}}-C_6H_5 \; \longrightarrow \; CH_3-\overset{O}{\underset{\|}{C}}-\overset{CH_3}{\underset{|}{N}}-C_6H_5 \; + \; CH_3-\overset{O}{\underset{\|}{C}}-OH$$

16.31 To determine what carboxylic acid and amine are needed to form an amide, merely break the nitrogen–carbonyl bond in the amide, add an –OH to the carbonyl to form the carboxylic acid and add a hydrogen to the nitrogen to form the amine; thus:

(a)

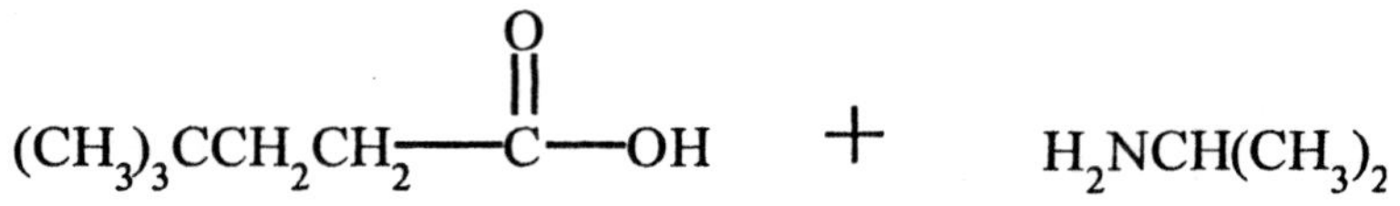

$$(CH_3)_3CCH_2CH_2-\overset{\overset{O}{\|}}{C}-OH \quad + \quad H_2NCH(CH_3)_2$$

(b)

$$2\ CH_3CH_2-\overset{\overset{O}{\|}}{C}-OH \quad + \quad H_2NCH_2CH_2CH_2NH_2$$

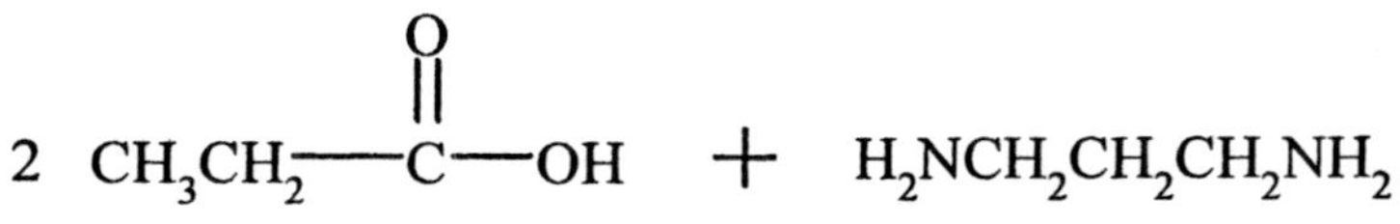

16.33 To determine what dicarboxylic acid and diamine are needed to form a polyamide, merely break the nitrogen–carbonyl bond in the two amide linkages of the polyamide. Then, add an –OH to the both carbonyls to form the dicarboxylic acid and add a hydrogen to both nitrogens to form the diamine; thus:

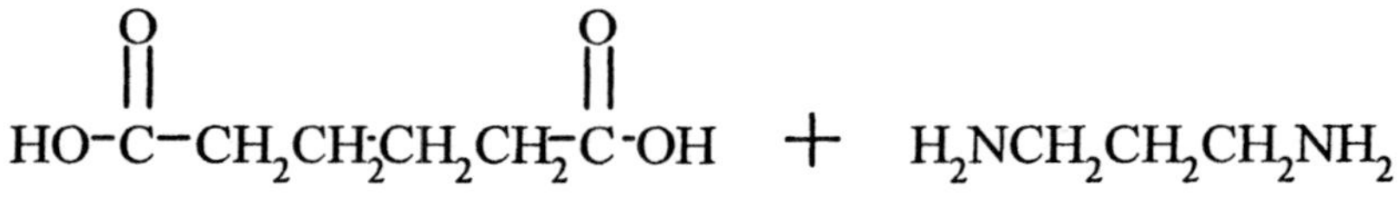

$$HO-\overset{\overset{O}{\|}}{C}-CH_2CH_2CH_2CH_2-\overset{\overset{O}{\|}}{C}-OH \quad + \quad H_2NCH_2CH_2CH_2NH_2$$

16.35 (a) N-*t*-Butylbutanamide: Butanamide means that the amide is derived from butanoic acid; thus there are three carbons bonded to the carbonyl. N-*t*-Butyl denotes that a tert-butyl group is bonded to the nitrogen of the amide; thus the structure is:

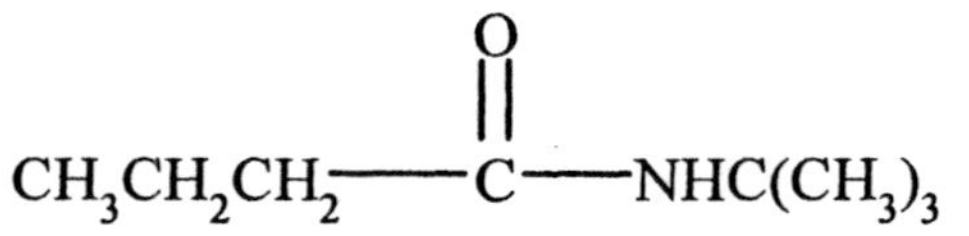

$$CH_3CH_2CH_2-\overset{\overset{O}{\|}}{C}-NHC(CH_3)_3$$

(b) N-Phenyl-3-methylhexanamide: Hexanamide means that the amide is derived from hexanoic acid, which means that a five-carbon chain is bonded to the carbonyl. The 3-methyl denotes that there is a methyl group on C-3 (where C-1 is the carbonyl carbon). Finally, N-phenyl denotes that a phenyl group is bonded to the amide nitrogen; thus the structure is:

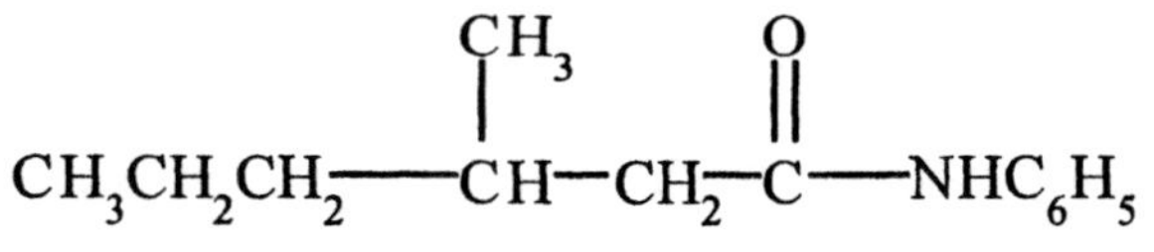

$$CH_3CH_2CH_2-\overset{\overset{CH_3}{|}}{CH}-CH_2-\overset{\overset{O}{\|}}{C}-NHC_6H_5$$

16.37 The first step in naming amides should be the determination of the carboxylic acid from which the amide derives. Replacing the carbonyl–nitrogen bond with a carboxylic acid group does so. The base name of the amide is formed from this carboxylic acid by replacing "oic acid" with "amide." Placing the name of the group preceded by "N" in front of the base name then specifies groups on the nitrogen.

(a) The carboxylic acid from which this compound derives is pentanoic acid; thus the base name is pentanamide. An isopropyl group is bonded to the nitrogen, which is denoted by placing N-isopropyl in front of pentanamide; thus this compound is N-isopropylpentanamide.

(b) The carboxylic acid from which this compound derives is propanoic acid; thus the base name is propanamide. A phenyl group is bonded to the nitrogen, which is denoted by placing N-phenyl in front of propanamide; thus this compound is N-phenylpropanamide.

(c) The carboxylic acid from which this compound derives is benzoic acid; thus the base name is benzamide. Two ethyl groups are bonded to the nitrogen, which is denoted by placing N,N-diethyl in front of benzamide; thus this compound is N, N-diethylbenzamide.

16.39 (a) Pentanamide has a hydrogen connected to the nitrogen, which can undergo hydrogen bonding to the carbonyl of another molecule. N, N-Dimethylpropanamide does not have a hydrogen connected to the nitrogen and thus cannot form hydrogen bonds between molecules. Thus pentanamide will have a higher boiling point than that of N, N-dimethylpropanamide.

(b) Both pentanamide and N, N-dimethylpropanamide can form hydrogen bonds to water through the carbonyl oxygen. Pentanamide can also hydrogen bond to water through the N–H, and thus the two compounds have very similar water solubilities, but pentanamide is slightly more soluble owing to the increased susceptibility to forming hydrogen bonds.

(c) The hydrogen bond between two amide molecules is stronger than the hydrogen bond between two carboxylic acid molecules; thus butanamide has a higher boiling point than that of butanoic acid.

(d) Both compounds can hydrogen bond with water, but the hydrogen bond formed by the amide group is slightly stronger than that formed by a carboxylic acid. Thus, the compounds have similar solubilities, but hexanamide is slightly more soluble.

(e) Both compounds can hydrogen bond with water, but the hydrogen bond formed by the amide group is slightly stronger than that formed by an alcohol. Thus, butanamide has the higher boiling point.

16.41 Basicity measures the ability of a compound to abstract a proton. Amines are more basic than water or an alcohol because the nitrogen of the amine has two nonbonded electrons that can bond to a proton. Amides are no more basic than alcohols or water, because the proximity of the carbonyl to the nitrogen limits the availability of the nonbonded electrons of the nitrogen. NaOH is the strongest base because the base is actually OH⁻, which seeks the positive charge of the proton to neutralize the anion. Thus:

$$\text{water} \approx \text{1-propanol} \approx \text{ethanamide} < \text{propylamine} < \text{NaOH}$$

16.43 (a) In the presence of an acid and water, an amide readily hydrolyzes (adds water) at O=C–N to form a carboxylic acid and an amine salt:

$$CH_3CH_2-\overset{O}{\underset{\|}{C}}-\overset{CH_3}{\underset{|}{N}}-CH_2CH_3 \xrightarrow[HCl]{H_2O} CH_3CH_2-\overset{O}{\underset{\|}{C}}-OH + H_2\overset{+}{N}-\underset{\underset{CH_3}{|}}{CH_2CH_3}\ Cl^-$$

(b) In the presence of a base and water, an amide readily hydrolyzes (adds water) at O=C–N to form a carboxylate salt and an amine:

$$C_6H_5-\overset{O}{\underset{\|}{C}}-\overset{H}{\underset{|}{N}}-C_6H_5 \xrightarrow{KOH} C_6H_5-\overset{O}{\underset{\|}{C}}-O^-\ K^+ + H_2N-C_6H_5$$

16.45 (a) Amides are classified as primary when two hydrogens are bonded to the nitrogen. Thus the three carbons must be part of the carbonyl as well as the hydrocarbon group bonded to the carbonyl. Thus the structure is:

$$CH_3CH_2-\overset{O}{\underset{\|}{C}}-NH_2$$

(b) A tertiary amide has no hydrogens bonded to the nitrogen. The fact that the compound has three methyl groups means that a methyl group must be bonded to the carbonyl and two other methyl groups must be bonded to the nitrogen. This accounts for the four carbons; thus the structure is:

$$CH_3-\overset{O}{\underset{\|}{C}}-\overset{CH_3}{\underset{|}{N}}-CH_3$$

(c) A lactam is a ring structure with an amide group in the ring. The exercise also states that the structure has five carbons that are all in the ring; thus:

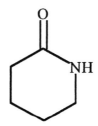

(d) A secondary amine has a nitrogen bonded to one hydrogen and two carbons. One of the groups to which the nitrogen is bonded must be a benzene ring, C_6H_5, which leaves CH_3 to be bonded to the nitrogen: $C_7H_9N - C_6H_5$ (of benzene ring) – NH (of amine) = CH_3. Thus the structure must be:

$$\text{C}_6\text{H}_5\text{-NHCH}_3$$

(e) A tertiary amine has a nitrogen bonded to three carbons. The *t*-butyl group accounts for C_4H_9 [= $C(CH_3)_3$] of the hydrocarbon structures that are bonded to the nitrogen. Thus there remains C_2H_6 to be distributed as two hydrocarbon groups that are bonded to the nitrogen; so the amine is tertiary. The two hydrocarbon groups must be $-CH_3$ groups; thus the structure is:

$$CH_3-N(CH_3)-C(CH_3)_3$$

16.47 (a) 5-Amino-3-methyl-2-pentanone: 2-Pentanone denotes that this compound is a ketone with a five-carbon chain and the carbonyl at C-2. 3-Methyl means that there is a methyl group on C-3, and 5-amino denotes that there is an amino (NH_2) group on C-5. The compound is not an amide, because the nitrogen is not bonded to the carbonyl.

$$CH_3-\underset{O}{\overset{\|}{C}}-\underset{CH_3}{\overset{|}{CH}}-CH_2CH_2NH_2$$

(b) 4-Pentene-1-amine: 4-Pentene denotes that the compound has a five-carbon chain with a double bond at C-4. 1-Amine at the end denotes that an amino group is bonded to C-1; thus:

$$H_2NCH_2CH_2CH_2CH=CH_2$$

16.49 (a) Atoms a, b, and c do not take part in double bonds and therefore all are sp^3 hybridized. Angles A and B involve only C–C and C–N bonds and are thus 109.5°.

(b) Atoms a, b, and c do not take part in double bonds and therefore they all are sp^3 hybridized. Angles A and B involve only C–C, N–H, and C–N bonds and are thus 109.5°.

16.51 (a) The hydrogen bond between two amide molecules is stronger than the hydrogen bond between two amine molecules; thus butanamide has a higher boiling point than that of hexanamine.

(b) Both compounds can hydrogen bond with water, but the hydrogen bond formed by the amide group is slightly stronger than that formed by an amine. Thus, the compounds have similar solubilities, but butanamide is slightly more soluble.

(c) The hydrogen bond between two amide molecules is stronger than the hydrogen bond between two carboxylic acid molecules; thus pentanamide has a higher boiling point than that of pentanoic acid.

(d) Both compounds can hydrogen bond with water, but the hydrogen bond formed by the amide group is slightly stronger than that formed by a carboxylic acid. Thus, the compounds have similar solubilities, but pentanamide is slightly more soluble.

(e) An amine salt is an ionic compound. Ionic interactions are much stronger than the hydrogen-bonding secondary forces of an amide; thus trimethylammonium chloride has a higher boiling point than that of hexanamide.

(f) A carboxylate salt is an ionic compound. Ionic interactions are much stronger than the hydrogen-bonding secondary forces of an amide; thus sodium butanoate has a higher boiling point than that of pentanamide.

(g) Both compounds have a single NH_2 group for every four carbons of the compound. The NH_2 groups can hydrogen bond to water; thus the two compounds have the same solubility in water.

16.53 An amine is a base that turns red litmus paper blue, whereas an amide does not. Thus, add a drop of the unknown to red litmus paper. If the paper turns blue, the unknown is an amine; if it does not, the unknown is an amide.

16.55 An amine is a base that turns red litmus paper blue, whereas an amide or an alcohol does not. Therefore, the unknown must be either propanol or ethanamide.

16.57 A polymer forms only if both reactants are bifunctional. In this exercise, only c has two bifunctional compounds that form a polymer:

$$HO-\overset{O}{\underset{\|}{C}}-(CH_2)_4-\overset{O}{\underset{\|}{C}}-OH \;+\; H_2N-(CH_2)_2NH_2 \;\xrightarrow{-H_2O}\; \left[-\overset{O}{\underset{\|}{C}}-(CH_2)_4-\overset{O}{\underset{\|}{C}}-\overset{H}{\underset{|}{N}}-(CH_2)_2\overset{H}{\underset{|}{N}}- \right]_n$$

16.59 (a) The first compound is an amide with five carbons, whereas the second compound is an amine and ketone with six carbons. In other words, in the first compound, the nitrogen is bonded to the carbonyl; whereas in the second, it is not; therefore, they are different compounds that are not isomers.

(b) Both compounds have the same molecular formula, $C_5H_{11}NO$, but the first compound is an amide, whereas the second compound is an amine and ketone; thus the two compounds are different compounds that are constitutional isomers.

(c) Both compounds have the same molecular formula C_3H_9N, but the first compound is a tertiary amine, whereas the second compound is a primary amine; thus these two compounds are different compounds that are constitutional isomers.

(d) Both compounds have the same molecular formula, C_5H_9NO, but the first compound is a five-membered heterocyclic amine, whereas the second compound is a six-membered heterocyclic amine; thus these two compounds are different compounds that are constitutional isomers.

16.61 $K_b = \dfrac{[OH^-][CH_3NH_3^+]}{[CH_3NH_2]} = 4.59 \times 10^{-4}$

$[OH^-] = [CH_3NH_3^+]$ $\qquad\qquad\qquad\qquad$ $[CH_3NH_2] = 1.0$

$[OH^-] = (4.59 \times 10^{-4})^{0.5} = 2.14 \times 10^{-2}$ $\qquad$ $[H_3O^+][OH^-] = 1 \times 10^{-14}$

$[H_3O^+] = 1 \times 10^{-14} / [OH^-] = 1 \times 10^{-14} / 2.14 \times 10^{-2} = 4.6 \times 10^{-13}$

pH = $-\log[H_3O^+]$ = 12.33

16.63 An amide will undergo hydrolysis to form a carboxylic acid and an amine. If the amine that is formed is ethylamine ($NH_2CH_2CH_3$), then the carboxylic acid must have three carbons in it (C_5 from amide – C_2 from amine). Therefore, the other compound is a carboxylic acid with three carbons: propanoic acid.

16.65 Ammonium chloride is an ionic compound that is not volatile. This lack of volatility results in a compound that does not have a strong odor.

Chapter 17

Stereoisomerism

17.1 Constitutional isomers are compounds that have the same molecular formula but different connectivity. Thus, isomers of C_4H_{10} are different compounds that differ in the way that the carbon atoms are bonded together. The most obvious first choice is to have them all bonded in a straight chain as $CH_3CH_2CH_2CH_3$. Next, the chain can be branched. The only possible branch point is on C-3; thus moving the C-1 methyl group to be a branch at C-3 gives $CH_3CH(CH_3)CH_3$. These two compounds are the only constitutional isomers of C_4H_{10}.

17.3 Butanol has the molecular formula $C_4H_{10}O$. An ether is a compound with formula R_1–O–R_2, where R_1, and R_2 are hydrocarbon structures. Possible combinations of R_1 and R_2 that can be formed from C_4H_{10} are $R_1 = C_2H_5$ and $R_2 = C_2H_5$; $R_1 = CH_3$ and $R_2 = C_3H_7$ (linear); $R_1 = CH_3$ and $R_2 = C_3H_7$ (branched); thus the isomers are $CH_3CH_2OCH_2CH_3$, $CH_3OCH_2CH_2CH_3$, and $CH_3OCH(CH_3)_2$.

17.5 *cis*-2-Butene is a compound with a double bond. The diastereomer of a compound with a double bond is a cis-trans isomer; thus the diastereomer of *cis*-2-butene is *trans*-2-butene:

$$\begin{array}{c} H \\ \diagdown \\ CH_3 \end{array} C = C \begin{array}{c} CH_3 \\ \diagup \\ H \end{array}$$

17.7 A chiral object is one that cannot be superimposed on its mirror image:

(a) A person cannot be superimposed on his or her mirror image; thus a person is chiral.

(b) An automobile cannot be superimposed on its mirror image; thus it is chiral.

(c) A basketball with no writing can be superimposed on its mirror image; thus it is achiral.

(d) The addition of a name written on a basketball results in an object that cannot be superimposed on its mirror image; thus it is chiral.

17.9 For a compound to be able to form enantiomers, it must contain a carbon atom that is bonded to four different structures.

(a) Every carbon has at least two hydrogens except C-3. C-3 is bonded to an –H, to –CH$_3$, and two –C$_2$H$_5$ groups. Therefore, no carbon in this molecule is bonded to four different groups; thus it cannot exist as a pair of enantiomers.

(b) Every carbon has at least two hydrogens except C-2. C-2 is bonded to an –H, to –C$_3$H$_7$, and two –CH$_3$ groups. Therefore, no carbon in this molecule is bonded to four different groups; thus it cannot exist as a pair of enantiomers.

(c) Every carbon has at least two hydrogens except C-2. C-2 is bonded to four different groups, –CH$_2$OH, –CH$_3$, –H, and –NH$_2$. Thus this compound can exist as two enantiomers, which are:

$$\begin{array}{cc}
\text{CH}_2\text{OH} & \text{CH}_2\text{OH} \\
| & | \\
\text{H—C*—NH}_2 & \text{H}_2\text{N—C*—H} \\
| & | \\
\text{CH}_3 & \text{CH}_3
\end{array}$$

(d) Every carbon either is part of a double bond (aromatic ring) or has at least two hydrogens. Therefore, no carbon in this molecule is bonded to four different groups; thus it cannot exist as a pair of enantiomers.

(e) Every carbon is either part of a double bond (aromatic ring) or has at least two hydrogens except the carbon that is connected to the phenyl ring. This carbon is bonded to four different groups, –C$_6$H$_5$, –CH$_3$, –H, and –CH$_2$NHCH$_3$. Thus this compound can exist as a pair of enantiomers, which are:

$$\begin{array}{cc}
\text{C}_6\text{H}_5 & \text{C}_6\text{H}_5 \\
| & | \\
\text{H—C*—CH}_2\text{NHCH}_3 & \text{CH}_3\text{NHCH}_2\text{—C*—H} \\
| & | \\
\text{CH}_3 & \text{CH}_3
\end{array}$$

17.11 In a comparison of two compounds, the first action should be to count the number of each type of atom to find a molecular formula. If the formulas are not the same, then the two compounds must be different compounds that are not isomers. Inspection of the compounds in this exercise shows that it is not true for any of the molecules in this exercise.

The next step should be to determine if the two molecules are the same molecule. In this exercise, this determination can best be made by superimposing the two molecules on top of each other. Care must be taken here to preserve the Fischer projection. In b, if either molecule is rotated 180° in the plane of the page, the other molecule is formed; that is the two molecules become superimposable. In c, the two molecules are superimposable as written; in e, the two molecules are superimposable as written.

Further inspection differentiates enantiomers, diastereomers, and constitutional isomers. In constitutional isomers, the connectivity of the atoms is different, whereas enantiomers or diastereomers have the same connectivity, yet the two molecules cannot be superimposed. Inspection shows that none of the compounds has different connectivity. Finally, for two molecules to be enantiomers, they must be mirror images of each other and have the same connectivity. In a and d, the two molecules can be seen as mirror images if the first compound is placed above the second.

Thus:

(a) 3; (b) 1; (c) 1; (d) 3; (e) 1

17.13 To identify D- and L-enantiomers, the compound must be written in the proper manner. The –H and –X groups must be in the horizontal and the two hydrocarbon substituents must be in the vertical plane in a Fischer projection. Moreover, R_1, the hydrocarbon with the fewest hydrogens, must be at the top of the projection and R_2, the compound with the most hydrogens, must be at the bottom. When the compound is written this way, a D-enantiomer has the hydrogen on the left, the L-enantiomer has it on the right.

(a) This Fischer projection is written correctly; thus it can be used to determine the D or L designation. This projection has the –H on the left side and is thus the D-enantiomer.

(b) This projection does not fit the aforegiven rules; thus it must be rotated 180° in the plane of the page to place the aldehyde group on top and the alcohol group on the bottom. When the compound is drawn in this way, the –H is on the right; thus it is the L-enantiomer.

(c) This Fischer projection is written correctly; thus it can be used to determine the D or L designation. This projection has the –H on the left side and is thus the D-enantiomer.

(d) This projection does not fit the aforegiven rules; thus it must be rotated 180° in the plane of the page to put the carboxylic acid group on top and the methyl group on the bottom. When the compound is drawn in this way, the –H is on the right; thus it is the L-enantiomer.

17.15 (a) Ethanol is not chiral and therefore will not rotate the plane of polarized light.

(b) D-Glucose is chiral and therefore will rotate the plane of polarized light. However, the D configuration does not specify in which direction it will rotate the plane, which must be experimentally determined.

(c) (+)-Phenylalinine is chiral and will rotate the plane of polarized light. The (+) denotes that it will rotate in the clockwise direction.

(d) A racemic mixture of glutamic acid is not chiral and will not rotate the plane of polarized light.

17.17 The D-enantiomer of a compound will rotate the plane of polarized light in the opposite direction and to the same extent as the L-enantiomer; thus the specific rotation of L-glutamic acid is +31.5°.

17.19 The (+) enantiomer of alanine must be the mirror image of (–)-alanine:

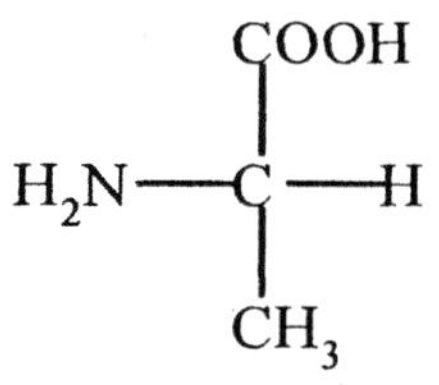

$$\begin{array}{c} COOH \\ | \\ H_2N-C-H \\ | \\ CH_3 \end{array}$$

17.21 The only chemical properties that chirality affects are optical rotation and chiral recognition. Thus, the water solubility of D-alanine at 25 °C is equal to the water solubility of L-alanine: 127 g/L.

17.23 The specific rotation, [α], can be determined by using the equation $[\alpha]=\dfrac{\alpha}{CL}$, where α is the observed rotation in degrees for a solution with concentration C (in g/mL) through a cell of length L (in dm = 10 cm). α = –3.6 °, C = 0.04 g/mL, L = 1 dm; thus [α] = –90°.

17.25 $[\alpha] = \dfrac{\alpha}{CL}$ → [α] = +223°, C = 0.06 g/mL, L = 1.5 dm. Thus α = [α]CL = +20.1°.

17.27 (a) In this compound, one carbon atom is bonded to four different groups, which gives the following pair of enantiomers, both of which are optically active:

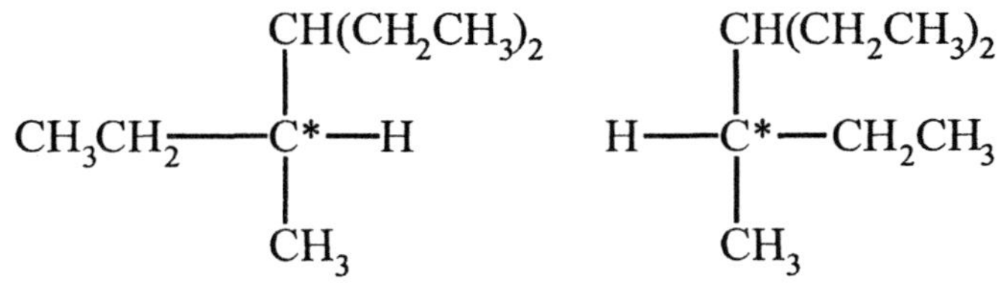

$$\begin{array}{cc} CH(CH_2CH_3)_2 & CH(CH_2CH_3)_2 \\ | & | \\ CH_3CH_2-C^*-H \qquad & H-C^*-CH_2CH_3 \\ | & | \\ CH_3 & CH_3 \end{array}$$

(b) Each of two carbons is bonded to four different groups. Therefore, this compound can form four stereoisomers. However, inspection of these structures shows that two of them (3 and 4) are equivalent and therefore this is a meso compound. Structures 1 and 2 are enantiomers and structure 3 is a meso compound. Structures 1 and 2 are diastereomers of structure 3 and vice versa. Structures 1 and 2 are optically active, but structure 3 is not.

```
      CH₂CH₃              CH₂CH₃              CH₂CH₃              CH₂CH₃
        |                    |                    |                    |
  H —— C*—— CH₃       CH₃—— C*—— H        CH₃—— C*—— H        CH₃—— C*—— H
        |                    |                    |                    |
  CH₃—— C*—— H        H —— C*—— CH₃       CH₃—— C*—— H    =   CH₃—— C*—— H
        |                    |                    |                    |
      CH₂CH₃              CH₂CH₃              CH₂CH₃              CH₂CH₃

        1                    2                    3                    4
```

(c) Each of two carbons is bonded to four different groups. Therefore, this compound can form four stereoisomers. Structures 1 and 2 are diastereomers of structures 3 and 4 and vice versa. Structures 1 and 2 are enantiomers, as are structures 3 and 4. All four compounds are optically active.

```
      CH₂CH₃              CH₂CH₃              CH₂CH₃              CH₂CH₃
        |                    |                    |                    |
  H —— C*—— OH       HO —— C*—— H        HO —— C*—— H        H —— C*—— OH
        |                    |                    |                    |
  CH₃—— C*—— H       H —— C*—— CH₃       CH₃—— C*—— H        H —— C*—— CH₃
        |                    |                    |                    |
      CH₂CH₃              CH₂CH₃              CH₂CH₃              CH₂CH₃

        1                    2                    3                    4
```

17.29 The differences in the boiling point and optical activity signify that these two compounds are not enantiomers. Therefore, the other two compounds must be the enantiomers of these two compounds. Structure 3 is the enantiomer of structure 1; structure 4 is the enantiomer of structure 2. Thus, the boiling point of structure 3 is the same as that of structure 1, 180–182°C, and its specific rotation is –15°. The boiling point of structure 4 is 163–165°C, the same as that of structure 2 and its specific rotation is –26°.

```
         CH₃                  CH₃
          |                    |
   HO —— C*—— H         HO —— C*—— H
          |                    |
   H₃C —— C*—— H         H —— C*—— CH₃
          |                    |
         CH₂OH                CH₂OH

           3                    4
```

17.31 (a) Each carbon that is bonded to four different groups is a tetrahedral stereocenter. There are four such carbons in this molecule:

$$\underset{}{\overset{O}{\overset{\|}{CH}}}-\overset{OH}{\underset{}{HC^*}}-\overset{OH}{\underset{}{HC^*}}-\overset{OH}{\underset{}{HC^*}}-\overset{OH}{\underset{}{HC^*}}-\overset{OH}{\underset{}{CH_2}}$$

A molecule with n tetrahedral stereocenters can form 2^n stereoisomers; thus this molecule can form 16 stereoisomers.

(b) Each carbon that is bonded to four different groups is a tetrahedral stereocenter. There are three such carbons in this molecule:

$$H_2N-\underset{CH_3}{C^*H}-\overset{O}{\overset{\|}{C}}-\underset{H}{N}-\underset{CH_2OH}{C^*H}-\overset{O}{\overset{\|}{C}}-\underset{H}{N}-\underset{CH_2COOH}{C^*H}-\overset{O}{\overset{\|}{C}}-OH$$

A molecule with n tetrahedral stereocenters can form 2^n stereoisomers; thus this molecule can form 8 stereoisomers.

17.33 A cyclic compound can exhibit chirality if a carbon in the ring has two different nonring substituents and the ring is not symmetric with respect to that carbon.

(a) 3-Methylcyclohexanone: C-3 has two different nonring substituents, and the ring is not symmetric at C-3. Therefore, there are two optically active enantiomers:

(b) 1,1Dimethylcyclopentane: No carbon in the ring has two different nonring substituents; each carbon has either two hydrogens or, in the case of C-1, two methyl groups. Thus this compound is not optically active.

(c) 2-Methyl-1, 3-cyclohexanedione: This compound has a carbon, C-2, that has two different nonring substituents. However, the ring is symmetric about this carbon; thus this compound is not optically active, and there are no stereoisomers.

(d) 1-Hydroxy-2-methylcyclohexane: This compound has two carbons, each of which has two different nonring substituents; thus there are four stereoisomers. Compounds 1 and 2 are a pair of trans enantiomers, whereas compounds 3 and 4 are a pair of cis enantiomers. All stereoisomers are optically active. Stereoisomers 1 and 2 are each diastereomers of compounds 3 and 4 and vice versa.

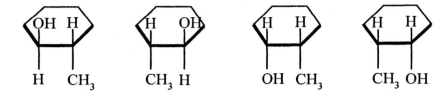

(e) 1,2-Dihydroxycyclohexane: This compound has two carbons, each of which has two different nonring substituents; thus there are four stereoisomers. However, inspection shows that two of the compounds are the same, the cis meso compound that is not optically active. The pair of trans compounds are enantiomers and are optically active.

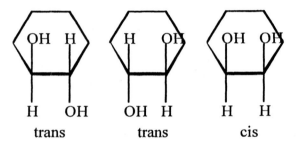

17.35 (a) Each carbon that is bonded to four different groups is a tetrahedral stereocenter. There are four such carbons in this molecule:

A molecule with n tetrahedral stereocenters can form 2^n stereoisomers; thus this molecule can form 16 stereoisomers.

(b) Each carbon that is bonded to four different groups is a tetrahedral stereocenter. There is one such carbon in this molecule:

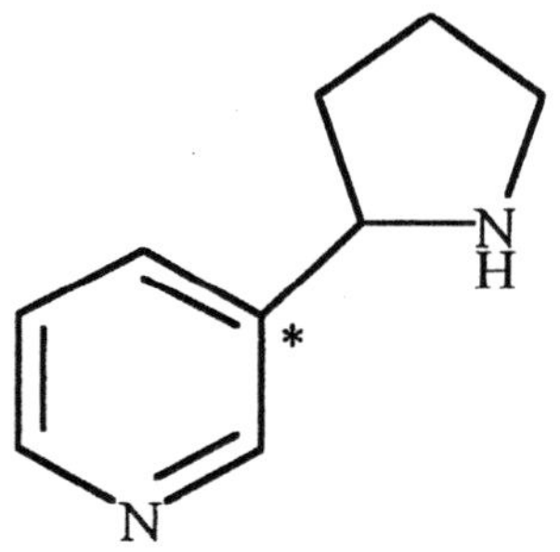

A molecule with n tetrahedral stereocenters can form 2^n stereoisomers; thus this molecule can form 2 stereoisomers.

(c) Each carbon that is bonded to four different groups is a tetrahedral stereocenter. There are six such carbons in this molecule:

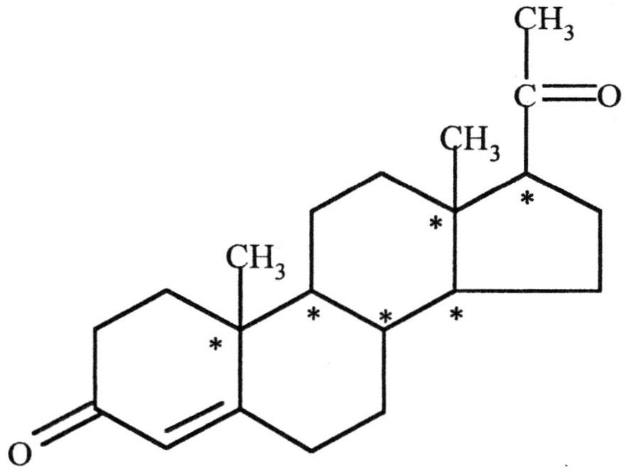

A molecule with n tetrahedral stereocenters can form 2^n stereoisomers; thus this molecule can form 64 stereoisomers.

17.37 In a comparison of two compounds, the first action should be to count the number of each type of atom to find a molecular formula. If the formulas are not the same, then the two compounds must be different compounds that are not isomers. Inspection of the compounds in this exercise shows that the two compounds in part o are different compounds ($C_7H_{13}Cl$ versus C_8H_{16}) that are not isomers of each other.

The next step should be to determine if the two molecules are the same molecule. In this exercise, this determination can best be made by superimposing the two molecules on top of each other. Care must be taken here to preserve the Fischer projections and it may simplify the comparison if both structures are written in a similar fashion. In i and j, if either molecule is rotated 180° in the plane of the page, the other molecule is formed; that is, the two molecules become superimposable. In b, e, and g, the two molecules are superimposable as written.

Further inspection differentiates enantiomers, diastereomers, and constitutional isomers. In constitutional isomers, the connectivity of the atoms is different, whereas enantiomers or diastereomers have the same connectivity, yet the two molecules cannot be superimposed. In a, d, and m, the two compounds have the same molecular formula but different connectivity and thus are constitutional isomers. Next, if two molecules are enantiomers, they must be mirror images of each other and have the same connectivity. In c, the two molecules can be seen as mirror images if the first compound is placed above the second. In h and l, the two molecules are mirror images if the mirror is placed between the two. Last, the compounds in f, k, and n have the same connectivity and yet are not mirror images of each other; thus they are diastereomers.

Thus:

(a) 2; (b) 1; (c) 3; (d) 2; (e) 1; (f) 4; (g) 1; (h) 3; (i) 1; (j) 1; (k) 4; (l) 3; (m) 2; (n) 4; (o) 5

17.39 4-Chloro-2-pentene has one tetrahedral stereocenter at C-3. However, owing to the double bond, there are two geometric stereoisomers (cis and trans). Thus this compound can form four diastereomers: a pair of cis enantiomers and a pair of trans enantiomers.

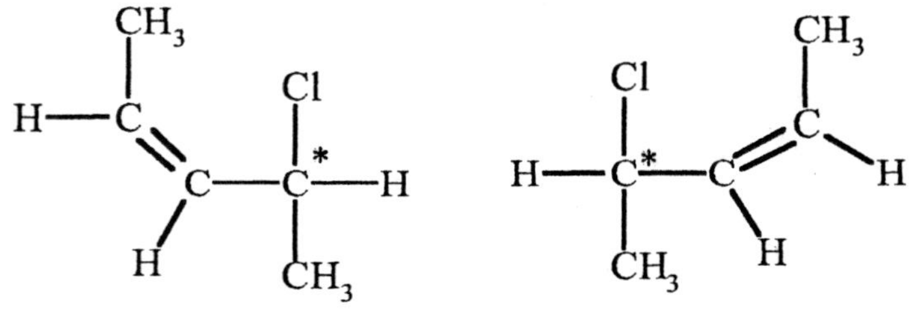

Cis pair of enantiomers

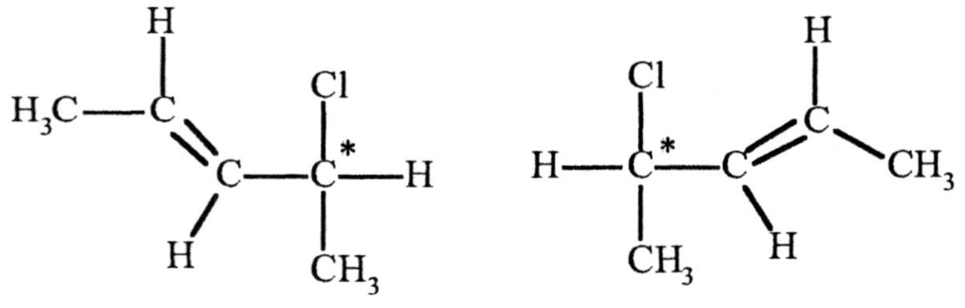

Trans pair of enantiomers

17.41 Both reactants, HCl and 1-butene, are achiral and therefore the reaction cannot incorporate chiral recognition. Thus, there is an equal probability that the Cl will add to

C-2 from either side of the double bond to form equal amounts of both enantiomers of the product. Therefore, the product is a racemic mixture of the two enantiomers of 2-chlorobutane that will not be optically active.

17.43 Both geometric isomers are achiral and therefore chiral recognition cannot be responsible for the difference in the physiological effect of the two isomers. A mechanism similar to chiral recognition in which a biological system preferentially recognizes one geometric isomer over the other is responsible for the different physiological response.

Chapter 18

Carbohydrates

18.1 (a) An aldose is a molecule with an aldehyde group and numerous hydroxy (–OH) groups. It is also called a polyhydroxyaldehyde.

(b) A hexose is an aldose or a ketose with six carbons. A ketose is a molecule with a ketone group and numerous hydroxy groups.

(c) A ketopentose is a ketose with five carbons. It may also be called a five-carbon polyhydroxyketone.

(d) An aldotetrose is an aldose with four carbons. It may also be called a four-carbon polyhydroxyaldehyde.

(e) All saccharides in the D family have the D configuration at the tetrahedral stereocenter farthest from the carbonyl carbon. In other words, the –OH group is on the right side of the carbon atom in the Fischer projection.

18.3 An aldopentose is a five-carbon molecule with hydroxy groups and an aldehyde. The structural formula that does not denote stereoisomerism is:

$$CH_2OH-\underset{\underset{OH}{|}}{CH}-\underset{\underset{OH}{|}}{CH}-\underset{\underset{OH}{|}}{CH}-\underset{\underset{O}{\|}}{CH}$$

Therefore, the carbon at one end is part of a carbonyl, the carbon at the other end has two hydrogens and an –OH bonded to it, and the other three carbons are tetrahedral stereocenters. The presence of three tetrahedral stereocenters means that there exist $2^3 = 8$ stereoisomers of this compound. Half of the stereoisomers are D and the other half are L; thus there are four L-aldopentoses.

18.5 A ketohexose is a six-carbon molecule with hydroxy groups and a ketone at C-2. Therefore, the carbon at each end of the molecule has two hydrogens and an –OH bonded to it, one carbon is part of the carbonyl, and the other three carbons are tetrahedral stereocenters. The presence of three tetrahedral stereocenters means that there exist $2^3 = 8$ stereoisomers of this compound, four of which are D-ketohexoses. On the other hand, an aldohexose is a six-carbon molecule with hydroxy groups and an aldehyde. The carbon at one end is part of a carbonyl, the carbon at the other end has two hydrogens and an –OH bonded to it, and the other four carbons are tetrahedral stereocenters. The presence of four tetrahedral stereocenters means that there exist $2^4 = 16$ stereoisomers of this compound. Half of the stereoisomers are D and the other half are L; thus there are eight D-aldohexoses.

18.7 (a) Glucose is an aldehyde, whereas sorbose is a ketone; therefore these two compounds cannot be stereoisomers. They both have six carbons and are thus constitutional isomers.

(b) Fructose and sorbose are both ketohexoses and are therefore stereoisomers of each other. However, they are not mirror images of each other; therefore they are diastereomers.

(c) (+)-Tagatose and (–)-tagatose denote two molecules that differ only in their stereo arrangement and are mirror images of each other; therefore they are enantiomers.

(d) D-Sorbose and L-sorbose denote two molecules that differ only in their stereo arrangement and are mirror images of each other; therefore they are enantiomers of each other.

(e) 2-Deoxy-D-ribose has one oxygen fewer than D-ribose, and therefore these two molecules are different compounds that are not isomers.

18.9 Haworth projections are used to represent cyclic structures of monosaccharides.

18.11 A pyranose ring is a six-membered cyclic hemiacetal ring.

18.13 The difference between the α and β anomers of D-glucose lies in the relative positions of the –CH$_2$OH on C-5 and the –OH on C-1. In the α-anomer, these two groups are situated trans to each other; whereas in the β-anomer, they have a cis relation.

18.15 The name allose denotes that the ring has five carbons and an oxygen and that the –OH groups on C-2, C-3, and C-4 are all on the same side of the ring, opposite to the –CH$_2$OH on C-5. D-Allose denotes that –CH$_2$OH on C-5 is above the ring in the Haworth projection. Finally, α-D-allose denotes that the –CH$_2$OH group on C-5 is trans to the –OH on C-1. Thus the structure is:

18.17 The name psicose denotes that the ring has four carbons and an oxygen and that the –OH groups on C-3 and C-4 are on the opposite side of the ring from the –CH$_2$OH on C-5. D-Psicose denotes that the –CH$_2$OH on C-5 is above the ring in the Haworth projection. Finally, β-D-psicose denotes that the –CH$_2$OH group on C-5 is cis to the –OH on C-2. Thus, the structure is:

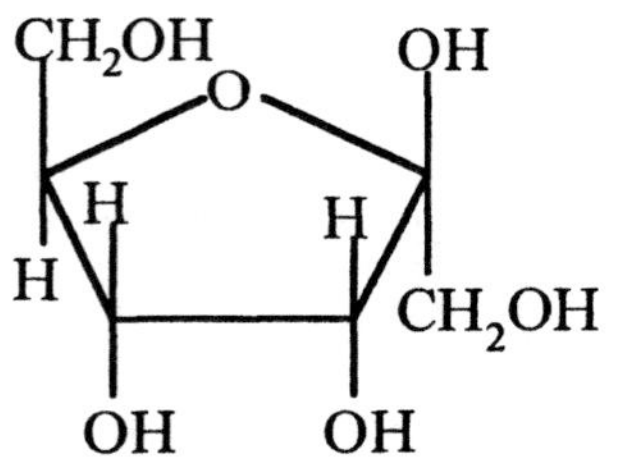

18.19 In solution, some of the cyclic acetal structure of β-D-glucose opens to form the linear aldehyde. The aldehyde structure is then present to give a positive Benedict's test. A positive test with Benedict's reagent is indicated by the disappearance of the blue color of Benedict's reagent and the formation of a red precipitate of Cu$_2$O.

18.21 Examination of Figures 18.1 and 18.2 shows that

(a) D-Glyceraldehyde has only one carbon that is a tetrahedral stereocenter.

(b) D-Sorbose has three carbons that are tetrahedral stereocenters.

(c) L-Mannose has four carbons that are tetrahedral stereocenters.

(d) 2-Deoxy-D-ribose has two carbons that are tetrahedral stereocenters. In a deoxy compound, one of the –OH groups is replaced with an –H; thus that carbon is no longer a stereocenter. Ribose has three carbons that are tetrahedral stereocenters.

18.23 The hemiacetal in glucose can undergo dehydration in the presence of an alcohol to form a glycoside:

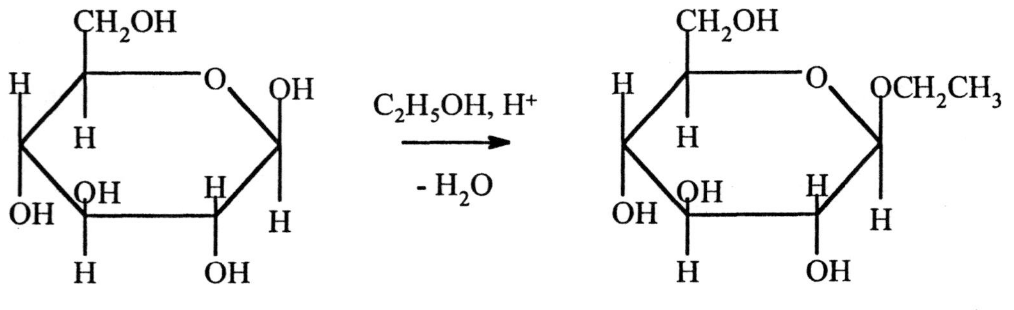

18.25 Reaction with Benedict's reagent results in an oxidation. In monosaccharides, the aldehyde undergoes oxidation to form a carboxylic acid. Therefore, D-galactose undergoes oxidation to form:

[structure of oxidized D-galactose ring showing CH₂OH, OH, H substituents and C=O with OH]

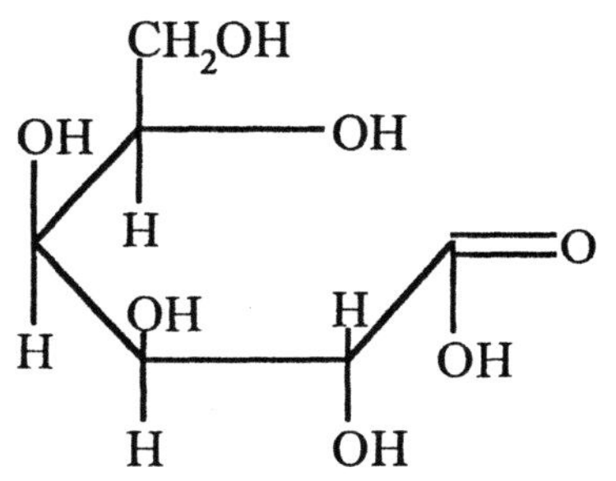

18.27 A reducing sugar is one that forms an open-chain structure that reacts with an oxidizing agent such as Benedict's reagent. To do so, it must have a hemiacetal linkage in the ring. Of the compounds listed in this exercise, all except b can undergo this reaction and are thus reducing agents and undergo mutarotation.

18.29 The α(1 → 4)-glycosidic linkage connects two aldohexose rings by creating a bond between the C-1 carbon in the α position of one ring and the C-4 position of the other.

18.31 (a) Maltose is two glucose rings bonded together by a glycosidic linkage. Digestion breaks this glycosidic linkage to leave two glucose rings.

(b) Lactose is a glucose ring bonded to a galactose ring by a glycosidic linkage. Digestion breaks this glycosidic linkage to leave a glucose and a galactose ring.

(c) Sucrose is a glucose ring bonded to a fructose ring by a glycosidic linkage. Digestion breaks this glycosidic linkage to leave a glucose and a fructose ring.

(d) Humans cannot digest cellobiose.

18.33 Figure 18.8 shows the structure of sucrose. It is:

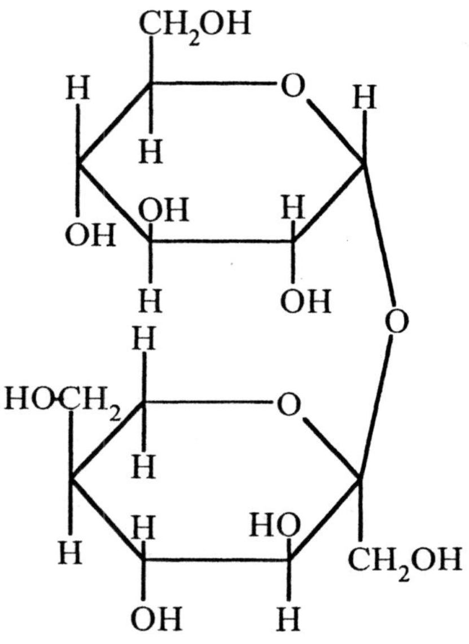

Sucrose is not a reducing sugar, because it does not contain a hemiacetal group that can open to the aldehyde structure and undergo oxidation.

18.35 The structure of two glucose rings linked by a β(1 → 6) linkage is:

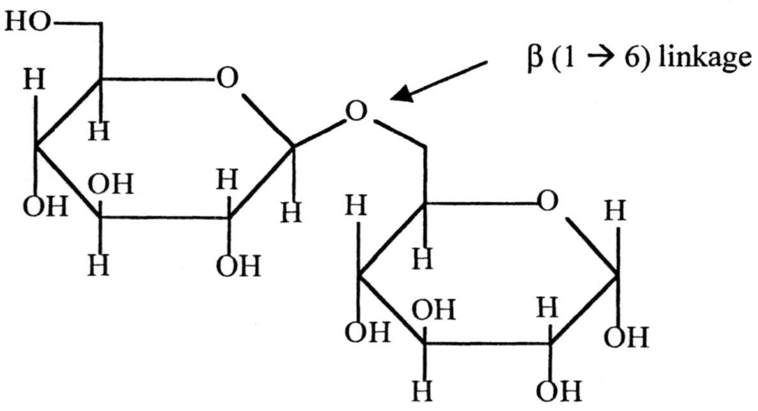

Gentiobiose is a reducing sugar because it contains a hemiacetal group in the ring that can open to the aldehyde structure and undergo oxidation.

18.37 Sucrose is not a reducing sugar and thus does not give a positive Benedict's test. Lactose is a reducing sugar and gives a positive Benedict's test. Because the unknown does not give a positive Benedict's test, it must be sucrose.

18.39 Both amylopectin and amylose are polysaccharides that have many D-glucose rings linked together by α(1 → 4) glycosidic linkages. However, amylose is a linear molecule, but amylopectin branches repeatedly by α(1 → 6) glycosidic linkages. Starch is a mixture of amylose and amylopectin that is found in plants as for the storage of D-glucose.

18.41 Humans can digest amylose, amylopectin, and glycogen to break the glycosidic linkages and yield D-glucose; humans cannot digest cellulose. Cows digest amylose, amylopectin, and cellulose to break the glycosidic linkages and yield D-glucose. Because they are herbivores and glycogens are present only in animals, grazing animals do not ingest glycogen.

18.43 N-Acetyl-D-glucosamine is a modified glucose unit. Chitin is this unit bonded by β(1 → 4) glycosidic linkages; thus the structure is:

18.45 Plants are the immediate source of organic compounds for animals. Photosynthesis is the ultimate source of organic compounds for animals because plants utilize photosynthesis to create organic compounds.

18.47 A carbohydrate is a compound that is an aldose or ketose. An aldose is a molecule with an aldehyde group and numerous hydroxy (–OH) groups. It is also called a polyhydroxyaldehyde. A ketose is a molecule with a ketone group and numerous hydroxy (–OH) groups. It is also called a polyhydroxyketone.

18.49 Only monosaccharides and disaccharides are called sugars.

18.51 In the Fischer projection, the placement of the –OH on the tetrahedral stereocenter farthest from the carbonyl (C-5) is on the right side in the D configuration. In the Haworth projection, the –CH$_2$OH on C-5 extends upward from the plane of the ring.

18.53 (a) Sucrose is two glucose rings bonded together and is therefore a disaccharide.

   (b) Galactose is a single aldohexose and is therefore a monosaccharide.

(c) Lactose is a glucose and a galactose ring linked together and is therefore a disaccharide.

(d) Amylose is a compound that has many D-glucose rings linked together by α(1 → 4) glycosidic linkages. Therefore, it is a polysaccharide.

(e) Amylopectin is a compound that has many D-glucose rings linked together by α(1 → 4) glycosidic linkages and branches repeatedly by α(1 → 6) glycosidic linkages. Therefore, it is a polysaccharide.

(f) Fructose is a single ketohexose and is therefore a monosaccharide.

(g) Cellulose is a linear molecule that has many D-glucose units connected together by β(1 → 4) glycosidic linkages. Therefore, it is a polysaccharide.

18.55 (a) α-D-Glucose and β-D-glucose denote two molecules that differ only in the way that the –OH on C-1 is bonded. Because all of the atoms are bonded in the same way in these compounds, they are stereoisomers of each other. However, they are not mirror images and are therefore diastereomers.

(b) Galactose and fructose are both hexoses, but fructose is a ketone, whereas galactose is an aldehyde; therefore they cannot be stereoisomers. They both have six carbons and thus are constitutional isomers

(c) (+)-Galactose and (–)-galactose denote two molecules that differ only in their stereo arrangement and are mirror images of each other; therefore they are enantiomers.

(d) Maltose and cellobiose are disaccharides that consist of two glucose molecules and therefore have the same molecular formula. However, different linkages in the two compounds bond the two rings. Therefore, the two compounds are not stereoisomers but are constitutional isomers.

(e) Amylose is a polysaccharide, whereas glucose is a monosaccharide. Therefore, amylose has many more carbon atoms than does glucose; therefore these two compounds are different compounds that are not isomers.

(f) Maltose is a disaccharide, whereas glucose is a monosaccharide. Therefore, maltose has twice as many carbon atoms as glucose has; therefore these two compounds are different compounds that are not isomers.

(g) Cellulose is a linear polysaccharide that has many D-glucose rings linked together by β(1 → 4) glycosidic linkages. Amylose also is a linear polysaccharide that has many D-glucose rings linked together by α(1 → 4) glycosidic linkages. Therefore, they have the same molecular formula but different connectivity and are therefore constitutional isomers.

18.57 Storage polysaccharides are compounds that have many units (usually D-glucose) linked together by glycosidic linkages. These linkages can be broken to yield the units such as D-glucose that can then be used to generate energy or synthesize other compounds. Structural polysaccharides, however, are used to construct cell walls in plants. These walls then impart stability and macroscopic shape to the plant.

18.59 Grazing animals such as cows have microorganisms in their digestive tracts that possess cellulase, the enzyme required to break the β(1 → 4) glycosidic linkages in cellulose. When the microorganism breaks these linkages, D-glucose remains, which the animals use for energy. Grazing animals themselves have amylase and maltase, the enzymes needed in digestion to break the α(1 → 4) glycosidic linkage of starch. This decomposition also results in glucose that the animal can use for energy.

18.61 Hydrolysis breaks the glycosidic linkages between monosaccharides; thus if 1 mol of carbohydrate yields 4 mol of monosaccharide, the carbohydrate must be made of four monosaccharides. In other words, it is a tetrasaccharide.

18.63 The structure of isomaltose is:

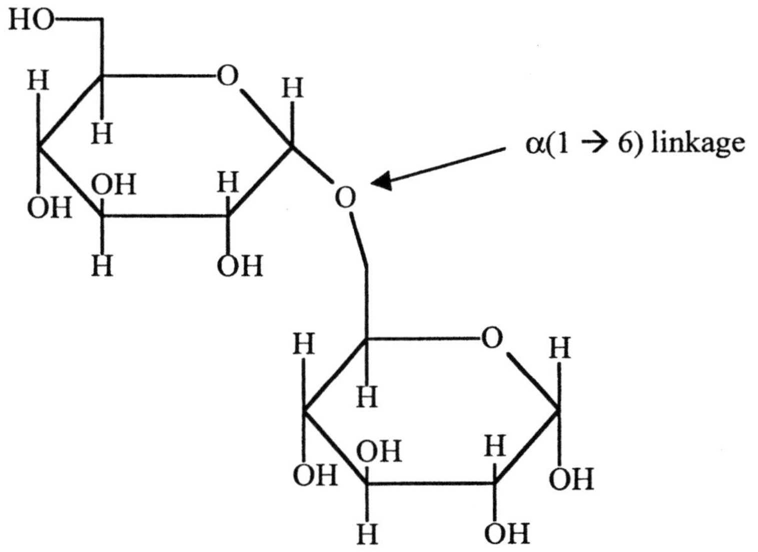

The presence of a hemiacetal group on the bottom ring in this structure means that isomaltose is a reducing sugar and undergoes mutarotation.

18.65 $6\ CO_2 + 6\ H_2O$ + sunlight + chlorophyll → $(CH_2O)_6 + 6\ O_2$

18.67 This is the cycle that oxygen undergoes as it is used and produced by plants. Plants produce oxygen as a result of photosynthesis. Plants and animals also use oxygen in metabolism to generate energy.

18.69 The glucose residue, $C_6H_{10}O_5$ has a molecular mass of 162 amu. The number of residues in a polysaccharide is the molecular mass of the polysaccharide divided by 162. Therefore, there are 1000 glucose residues in either amylose or cellulose of molecular mass 162,000.

18.71 This stereospecific linkage must be formed by chiral recognition whereby an enzyme allows only the two correct stereoisomers to come together in a specified manner to form the correct glycosidic linkage.

18.73 Lactose intolerance is a result of the absence of the proper enzyme, lactase, needed to digest lactose. Symptoms of lactose intolerance include abdominal distention and cramping, nausea, pain, and diarrhea. These symptoms are temporary, however, and will subside in the absence of lactose. Galactosemia has long-term adverse effects if not detected early in infancy. Long-term adverse effects include mental retardation, impaired liver function, cataracts, and death.

Chapter 19

Lipids

19.1   The fatty acids found in animals and plants have three particular characteristics: they have an even number of carbons; are unbranched; and, when double bonds are present, have a cis configuration. In this exercise, only compound 1 fulfills all of these requirements. Compound 2 is branched, whereas compound 3 has an odd number of carbons.

19.3   Saturated fatty acids have no carbon–carbon double bonds. An unsaturated fatty acid contains one or more carbon–carbon double bond in the cis configuration. Unsaturated fatty acids are further broken up into monosaturated and polyunsaturated fatty acids. A monosaturated fatty acid contains only one carbon–carbon double bond, whereas a polyunsaturated fatty acid contains more than one carbon–carbon double bond.

19.5   Compound 2 has a higher melting point because molecules of compound 2 can pack more tightly together owing to the flexible linear chain of single bonds. The double bond in compound 1 causes the chain to kink and thus the chains cannot pack as tightly as the saturated chain can. The closer packing of compound 2 results in stronger intermolecular interactions that result in a higher melting point.

19.7   A triacylglycerol is the triester of glycerol. Lauric acid is a saturated acid with 12 carbons, myristic acid is a saturated acid with 14 carbons, and oleic acid is a monosaturated acid with 18 acids. The triacylglycerol is created by removing the –OH from the acid and an –H from a hydroxyl group on glycerol and, then creating a bond between the carbonyl of the acid to the –O– of glycerol. Thus this compound is:

$$\begin{array}{l} CH_2-O-\overset{\overset{O}{\|}}{C}-(CH_2)_{10}CH_3 \\ |\\ CH-O-\overset{\overset{O}{\|}}{C}-(CH_2)_{12}CH_3 \\ |\\ CH_2-O-\overset{\overset{O}{\|}}{C}-(CH_2)_7-CH=CH(CH_2)_7CH_3 \end{array}$$

19.9   Melting point decreases as the amount of double bonds in the triacylglycerol increases. Compound a has no double bonds, compound b has three double bonds, and compound c has one double bond. Thus, in increasing order of melting point, compound b < compound c < compound a.

19.11 (a) Saponification is the reaction of the ester group with water in the presence of a base. The result is that the ester cleaves to form the alcohol and the carboxylate salt. In triacylglycerols, this reaction takes place at each ester to form:

$$\begin{array}{l} CH_2-O-\overset{O}{\overset{\|}{C}}-(CH_2)_{12}CH_3 \\ | \quad\quad O \\ \quad\quad \| \\ CH-O-C-(CH_2)_{14}CH_3 \\ | \quad\quad O \\ \quad\quad \| \\ CH_2-O-C-(CH_2)_7-CH=CH(CH_2)_7CH_3 \end{array} \xrightarrow[NaOH]{H_2O} \begin{array}{l} CH_2-OH \\ | \\ CH-OH \\ | \\ CH_2-OH \end{array} + \begin{array}{l} \overset{O}{\overset{\|}{NaO-C}}-(CH_2)_{12}CH_3 \\ \overset{O}{\overset{\|}{NaO-C}}-(CH_2)_{14}CH_3 \\ \overset{O}{\overset{\|}{NaO-C}}-(CH_2)_7-CH=CH(CH_2)_7CH_3 \end{array}$$

(b) Catalytic hydrogenation in the presence of Pt results in the addition of H_2 across the carbon–carbon double bond. The only part of the molecule that this can take place in is the bottom fatty acid:

$$\begin{array}{l} CH_2-O-\overset{O}{\overset{\|}{C}}-(CH_2)_{12}CH_3 \\ | \quad\quad O \\ \quad\quad \| \\ CH-O-C-(CH_2)_{14}CH_3 \\ | \quad\quad O \\ \quad\quad \| \\ CH_2-O-C-(CH_2)_7-CH=CH(CH_2)_7CH_3 \end{array} \xrightarrow[Pt]{H_2} \begin{array}{l} CH_2-O-\overset{O}{\overset{\|}{C}}-(CH_2)_{12}CH_3 \\ | \quad\quad O \\ \quad\quad \| \\ CH-O-C-(CH_2)_{14}CH_3 \\ | \quad\quad O \\ \quad\quad \| \\ CH_2-O-C-(CH_2)_{16}CH_3 \end{array}$$

19.13 Bacterial hydrolysis of this compound breaks the ester linkages to form fatty acids and glycerol. One of the fatty acids that result is butanoic acid, $C_4H_8O_2$. Because of its low molar mass, butanoic acid is volatile and produces a foul odor.

$$\begin{array}{l} CH_2-O-\overset{O}{\overset{\|}{C}}-(CH_2)_2CH_3 \\ | \quad\quad O \\ \quad\quad \| \\ CH-O-C-(CH_2)_{14}CH_3 \\ | \quad\quad O \\ \quad\quad \| \\ CH_2-O-C-(CH_2)_7-CH=CH(CH_2)_5CH_3 \end{array} \xrightarrow{H_2O} \begin{array}{l} CH_2-OH \\ | \\ CH-OH \\ | \\ CH_2-OH \end{array} + \begin{array}{l} \overset{O}{\overset{\|}{HO-C}}-(CH_2)_2CH_3 \\ \overset{O}{\overset{\|}{HO-C}}-(CH_2)_{14}CH_3 \\ \overset{O}{\overset{\|}{HO-C}}-(CH_2)_7-CH=CH(CH_2)_5CH_3 \end{array}$$

19.15 By hydrogenation of the carbon–carbon double bonds in corn oil in the presence of a metal catalyst, the fatty acids become more saturated and thus linear. These linear chains pack more readily, which increases the melting point of the compound.

19.17 The waxes found in nature tend to be an ester that is the combination of a carboxylic acid and an alcohol. The acid and alcohol normally contain between 14 and 36 carbons each, are unbranched, and have an even number of carbons. Therefore, the structure of the wax should have an even number of carbons, between 14 and 36, on both sides of the oxygen that is next to the carbonyl and be linear. Compound 1 has only four carbons to the left of the carbonyl, whereas compound 3 has an odd number of carbons on both sides of the oxygen; thus compound 2 is the correct compound.

19.19 Both triacylglycerols and glycerophospholipids are created by forming ester linkages to the three –OH groups of glycerol. In triacylglycerol, all three esters are bonded to fatty acids. However, in glycerophospholipids, two of the ester links are bonded to fatty acids, but the third is linked to a phosphodiester group.

19.21 If the hydrolysis yields myristic acid and oleic acid, these two fatty acids must be bonded to two of the ester groups of glycerol. The presence of phosphoric acid and serine in the products tells us that the phosphodiester is bonded to serine (from Table 19.3). Thus, the structure is:

$$\begin{array}{l} CH_2-O-\overset{O}{\underset{\|}{C}}-(CH_2)_{12}CH_3 \\ | \quad\quad\quad O \\ CH-O-\overset{\|}{C}-(CH_2)_7-CH=CH(CH_2)_7CH_3 \\ | \quad\quad\quad O \\ CH_2-O-\overset{\|}{\underset{|}{P}}-O-CH_2CH-NH_3^+ \\ \quad\quad\quad\;\; O^- \quad\quad COO^- \end{array}$$

19.23 The hydrolysis breaks the top two ester groups to form –OH on glycerol and two fatty acids. The phosphodiester decomposes to phosphoric acid and the alcohol HOCH$_2$CH$_2$NH$_3^+$:

$$\begin{array}{ll} CH_2-OH & HO-\overset{O}{\underset{\|}{C}}-(CH_2)_{14}CH_3 \\ | & \quad\quad O \\ CH-OH & HO-\overset{\|}{C}-(CH_2)_{16}CH_3 \\ | & \quad\quad O \\ CH_2-OH & HO-\overset{\|}{\underset{|}{P}}-OH \quad\quad HO-CH_2CH_2NH_3^+ \\ & \quad\quad\; O^- \end{array}$$

19.25 Linoleic acid is a polyunsaturated acid with 18 carbons and choline is shown in Table 19.3. Linoleic acid will combine at the amine in sphingosine to form an amide group, whereas the phosphoric acid and choline combine at the –OH of sphingosine to form:

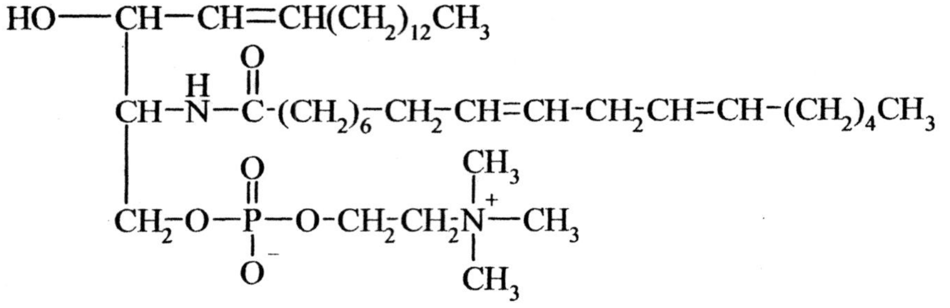

HO—CH—CH=CH(CH$_2$)$_{12}$CH$_3$
|
CH—N—C-(CH$_2$)$_6$—CH$_2$—CH=CH-CH$_2$CH=CH-(CH$_2$)$_4$CH$_3$
|
CH$_2$-O—P—O-CH$_2$CH$_2$N$^+$—CH$_3$
              |
              CH$_3$

19.27 An acid-catalyzed hydrolysis of this lipid breaks the amide and ester linkages to form sphingosine, oleic acid, phosphoric acid, and choline:

HO—CH—CH=CH(CH$_2$)$_{12}$CH$_3$ HO-C-(CH$_2$)$_7$—CH=CH(CH$_2$)$_7$CH$_3$
|
CH—NH$_2$
|
CH$_2$-OH   HO—P—OH   HO·CH$_2$CH$_2$N$^+$—CH$_3$
                   | |
                   OH   CH$_3$

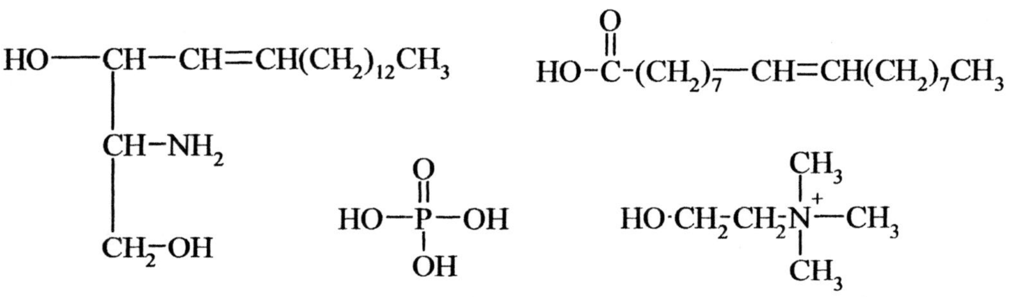

19.29 Steroids do not contain functional groups such as an ester or amide group that can be cleaved in hydrolysis to form smaller molecules.

19.31 A tetrahedral stereocenter is one that is bonded to four different groups. In testosterone, there are six such carbons, denoted by asterisks below. Because there are six tetrahedral stereocenters, there are $2^6 = 64$ stereoisomers of testosterone.

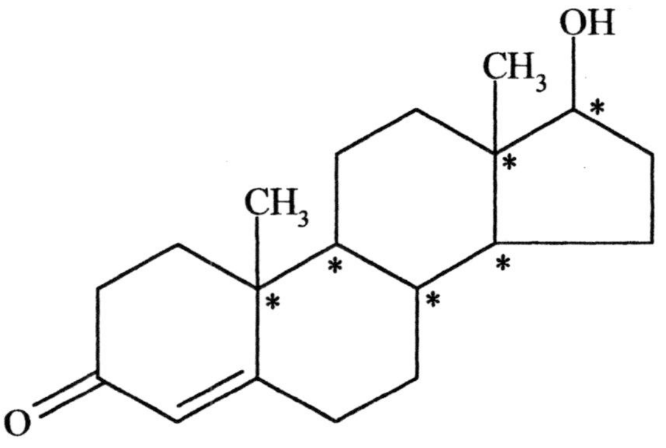

19.33 The sex hormones regulate the development of the sex organs and the production of sperm and ova. They also regulate the development of secondary sex characteristics: lack of facial hair, increased breast size, and high voice in women; facial hair, increased musculature, and deep voice in men.

19.35 Cholesterol is the precursor to steroid hormones and bile salts.

19.37 Bile salts are compounds that replace a carboxylic acid group that is bonded to the ring structure in a steroid with a carboxylate salt.

19.39 Hydrolysis can take place at the amide group. However, the part of the molecule that is removed is small compared with the overall structure of leukotriene D_4. Thus this change does not substantially alter the function of the molecule.

19.41 In a leukotriene, the 20-carbon chain of arachidonic acid and the carboxyl group remain intact. In a prostaglandin, a cyclopentane ring is formed between C-8 and C-12 of the 20-carbon chain.

19.43 Hormones are synthesized in and secreted by endocrine glands and then transported in the bloodstream to target tissues where they regulate the functions of cells. Eicosanoids are called local hormones because they are not transported in the blood stream but act in the tissue in which they were synthesized.

19.45 Vitamins are organic compounds that are required in trace amounts for normal metabolism but are not synthesized by the organism that requires them. They; therefore, must be included in diet.

19.47 Fat-soluble vitamins dissolve in fat but not water; water-soluble vitamins dissolve in water but not fat.

19.49 A complete description of these functions can be found in Table 19.4.

Vitamin A: plays a key role in vision by aiding proper function of mucous membranes.

Vitamin D: Regulates calcium and phosphate use and deposition in bone and cartilage.

Vitamin E: Acts as an antioxidant to protect cell membrane lipids from cleavage by O_2.

Vitamin K: Regulates formation of prothrombin, which is needed for blood clotting.

19.51 Membranes surround all cells and organelles.

19.53 Lipids with hydrophilic and hydrophobic parts, such as glycerophospholipids, sphingolipids, or cholesterol, form the membrane in the lipid bilayer.

19.55 Unsaturated fatty acids have cis double bonds that produce kinks in the chains. The bent chains result in a lipid bilayer that is not tightly packed, has weaker secondary forces, and has more flexibility.

19.57 The ions and D-glucose are hydrophilic (they like water). Therefore, the hydrophobic regions of the lipid bilayer repel them.

19.59 Active transport is the diffusion of a compound from an area of low concentration to an area of high concentration. This transport against a concentration gradient can happen only if energy is added to the system.

19.61 Because lactose is being transported from a region of low concentration to one of high concentration, this must be active transport.

19.63 Lipids are not very polar; therefore, the most polar molecule, CH_3OH, would be the least effective in dissolving lipids.

19.65 (a) This compound does not have a carbon–carbon double bond and therefore it does not have an ω number.

(b) The first double bond is seven carbons from the methyl end group of this acid; thus it is ω-7.

(c) The first double bond is six carbons from the methyl end group of this acid; thus it is ω-6.

19.67 Waxes form a protective coating on plants and animals. They protect against parasites, excessive water loss, and mechanical damage, and they waterproof fowl.

19.69 Digestion does not result in the cleavage of all the hydrolyzable groups. A mixture of products is usually obtained. For instance, triacylglycerols usually yield monoacylglycerols and fatty acids.

19.71 A lipid is hydrolyzable if it contains an ester or amide group that can be cleaved in the presence of water. Of the compounds listed here, only (a) sphingolipids, (c) glycerophospholipids, (f) triacylglycerols, and (g) waxes contain these groups and can undergo hydrolysis.

19.73 A lipid can undergo saponification if it contains an ester group that can be cleaved in the presence of a base and water. Of the compounds listed here, only (a) sphingolipids, (c) glycerophospholipids, (f) triacylglycerols, and (g) waxes contain esters and can undergo saponification.

Lipids 155

19.75 The three constitutional isomers differ in the position at which the fatty acids are bonded to the glycerol; thus they are:

$$\begin{array}{lll}
CH_2\text{-O-}\overset{O}{\underset{\|}{C}}\text{-}(CH_2)_{12}CH_3 & CH_2\text{-O-}\overset{O}{\underset{\|}{C}}\text{-}(CH_2)_{12}CH_3 & CH_2\text{-O-}\overset{O}{\underset{\|}{C}}\text{-}(CH_2)_{14}CH_3 \\
CH\text{-O-}\overset{O}{\underset{\|}{C}}\text{-}(CH_2)_{14}CH_3 & CH\text{-O-}\overset{O}{\underset{\|}{C}}\text{-}(CH_2)_{16}CH_3 & CH\text{-O-}\overset{O}{\underset{\|}{C}}\text{-}(CH_2)_{12}CH_3 \\
CH_2\text{-O-}\overset{O}{\underset{\|}{C}}\text{-}(CH_2)_{16}CH_3 & CH_2\text{-O-}\overset{O}{\underset{\|}{C}}\text{-}(CH_2)_{14}CH_3 & CH_2\text{-O-}\overset{O}{\underset{\|}{C}}\text{-}(CH_2)_{16}CH_3
\end{array}$$

19.77 The carbon–carbon double bond of oleic acid in triacylglycerol B yields a carboxylic acid when it undergoes air oxidation. This carboxylic acid has a bad odor and has a low enough molar mass to be volatile. Triacylglycerol A, however, does not have a carbon–carbon double bond and therefore will not undergo air oxidation.

19.79 A tetrahedral stereocenter is one that is bonded to four different groups. In progesterone there are six such carbons, denoted by asterisks below. Because there are six tetrahedral stereocenters, there are $2^6 = 64$ stereoisomers of progesterone.

19.81 Triacylglycerols are completely hydrophobic; they have no parts that like water, or are hydrophilic.

19.83 The fact that the protons are diffusing from an area of low concentration to high concentration means that this is active transport.

19.85 Coconut oil has fatty acids with very few double bonds. This usually results in a high melting point because the chains can pack well. However, the fatty acids in coconut oil have low molecular mass, usually fewer than 14 carbons per fatty acid. Thus, coconut oil is liquid at room temperature even though it is mostly saturated because it has a low melting point as a result of the low molecular mass of the fatty acids.

19.87 The ester that is formed is created by the reaction of the acid group of linoleic acid and the –OH group of cholesterol to form:

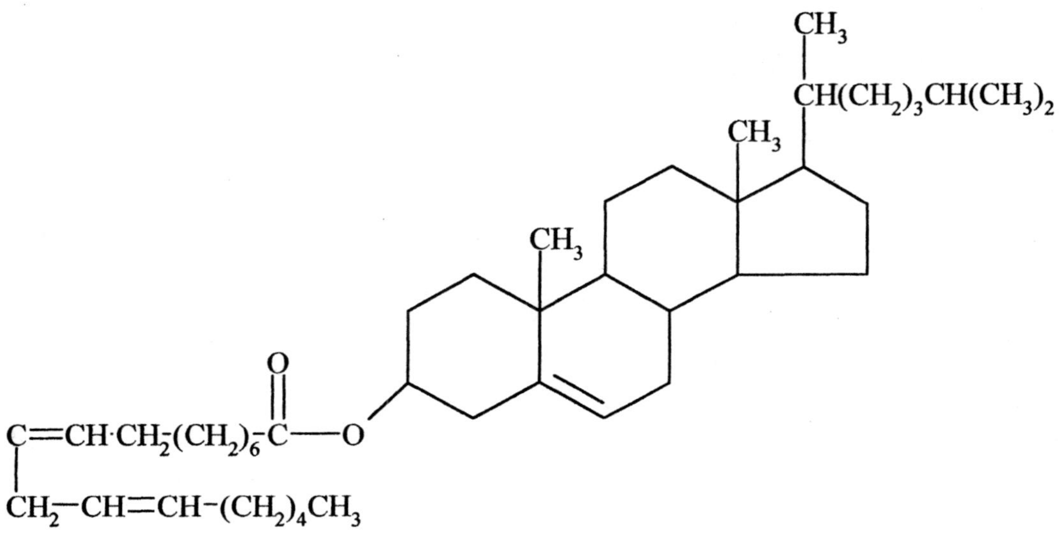

[structure of cholesteryl linoleate: cholesterol steroid ring system with CH$_3$ groups and CH(CH$_2$)$_3$CH(CH$_3$)$_2$ side chain, esterified at the 3-OH position with linoleic acid: C=CH·CH$_2$(CH$_2$)$_6$C(=O)—O— with CH$_2$—CH=CH—(CH$_2$)$_4$CH$_3$ branch]

19.89 NaOH will react with the triacylglycerols in the fats to form glycerol and sodium carboxylate salts. Glycerol is soluble in water owing to hydrogen bonding between H$_2$O and the –OH groups, whereas sodium carboxylate salts are soluble in water owing to the ionic nature of the compound. This increased solubility in water allows the clog to be washed out.

19.91 In biological membranes on Earth, the lipids organize to create an interior that is hydrophobic and an exterior that is hydrophilic, owing to the presence of water throughout the body. On Gibo, we can assume that heptane is present everywhere in the body, and thus the membrane must organize to create an interior that is hydrophilic and an exterior that is hydrophobic. This can be accomplished by organizing the same lipids in a reverse bilayer so that the hydrophobic part is on the outside of the cell and the hydrophilic part is on the inside.

Chapter 20

Proteins

20.1 An α-amino acid is one that has the nitrogen bonded to the carbon that is right next to the carbonyl; a β–amino acid is one that has the nitrogen bonded to the carbon that is two carbons from the carbonyl; and a γ-amino acid is one that has the nitrogen bonded to the carbon that is three carbons from the carbonyl. Thus:

Compound 1 is β; compound 2 is α; compound 3 is γ

20.3 (a) Gly; (b) Ser, Thr; (c) Phe, Tyr, Trp; (d) Ile, Thr; (e) Trp, His; (f) Met, Cys; (g) Asp, Glu; (h) Pro

20.5 The carboxylic acid group in the amino acid reacts with the amino group in an acid-base reaction to form a positively charged amino group and a negatively charged carboxyl group.

20.7 (a) At its isoelectric point, alanine exists as a zwitterion. At pH < 1, the solution is very acidic. Under this condition, the negatively charged carboxyl group reacts with free protons, which results in an amino acid that has a net positive charge. At pH > 12, the solution is very basic, and the positively charged amino group will give up a proton to leave an amino acid that that has a net negative charge. Thus:

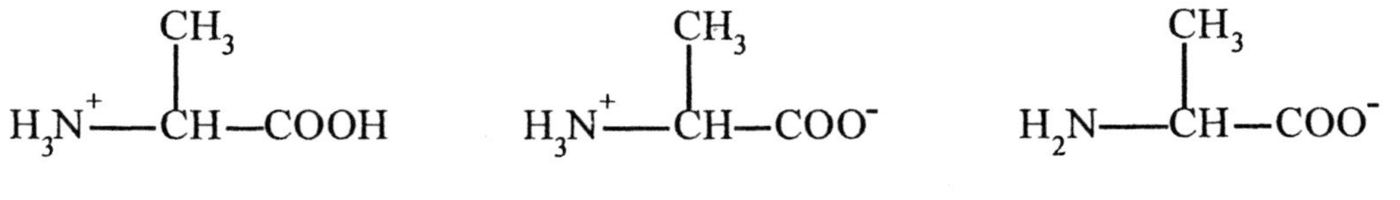

$$H_3N^+-CH(CH_3)-COOH \qquad H_3N^+-CH(CH_3)-COO^- \qquad H_2N-CH(CH_3)-COO^-$$

$$pH < 1 \qquad\qquad pH \sim pI\ and\ 7 \qquad\qquad pH > 12$$

(b) Alanine is a neutral compound at pI but is an ion at higher and lower pH. Ions are more soluble in water than are neutral compounds; thus alanine is least soluble in water at pI = 6.01. It will be most soluble at very high and very low pH.

(c) At pI, the molecule is neutral and does not migrate in electrophoresis. At pH = 7, the pH is within 2 of pI, and thus alanine does not migrate in electrophoresis.

20.9 • At pH = 7, the carboxylate and amino groups are charged.

- To name peptides, (i) the C-terminal residue keeps its amino acid name, (ii) the -ine or -ic acid of the other amino acid residues is replaced with –yl, and (iii) naming begins at the N-terminal residue.

Thus, Ser-Met is serylmethionine:

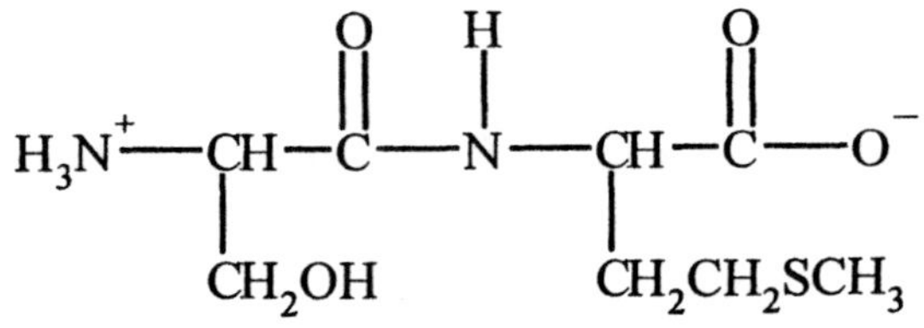

$$H_3N^+-CH(CH_2OH)-C(=O)-NH-CH(CH_2CH_2SCH_3)-C(=O)-O^-$$

20.11 • At pH = 7, the carboxylate and amino groups are charged.

- To name peptides, (i) the C-terminal residue keeps its amino acid name, (ii) the -ine or -ic acid of the other amino acid residues is replaced with –yl, and (iii) naming begins at the N-terminal residue.

Thus, Thr-Ala-Asp is threonylalanylaspartic acid:

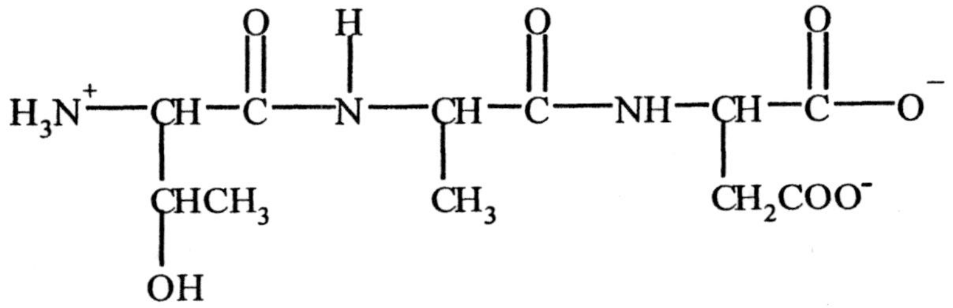

$$H_3N^+-CH(CH(OH)CH_3)-C(=O)-NH-CH(CH_3)-C(=O)-NH-CH(CH_2COO^-)-C(=O)-O^-$$

20.13 • At pH = 7, the carboxylate and amino groups are charged.

- To name peptides, (i) the C-terminal residue keeps its amino acid name, (ii) the -ine or -ic acid of the other amino acid residues is replaced with –yl, and (iii) naming begins at the N-terminal residue.

Thus, Gly-Pro-Ala is glycylprolylalanine:

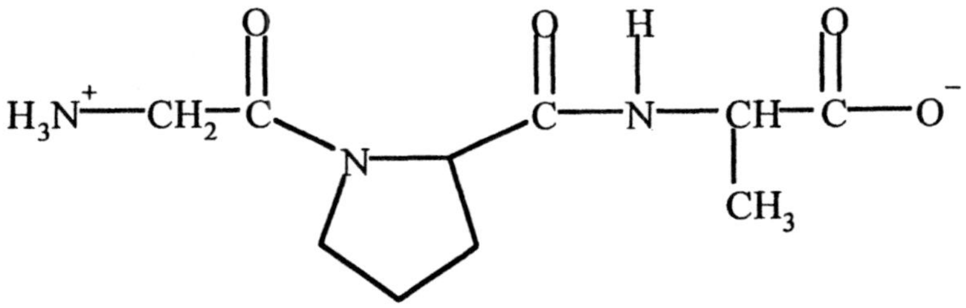

$$H_3N^+-CH_2-C(=O)-[Pro]-C(=O)-NH-CH(CH_3)-C(=O)-O^-$$

20.15 Constitutional isomers of a tripeptide exist as molecules with different connectivities of the three amino acids. For threonine, cysteine, and leucine, there are six constitutional isomers: Thr-Leu-Cys, Thr-Cys-Leu, Cys-Leu-Thr, Cys-Thr-Leu, Leu-Cys-Thr, and Leu-Thr-Cys.

20.17 Each residue can exist as a D- or L-enantiomer. Thus, the six possible stereoisomers are different combinations of these enantiomers: L-Lys-L-Ala-L-Glu, L-Lys-L-Ala-D-Glu, L-Lys-D-Ala-D-Glu, D-Lys-D-Ala-D-Glu, D-Lys-L-Ala-L-Glu, and D-Lys-D-Ala-L-Glu. Only the stereoisomer with all L residues would be found in nature.

20.19 Aspartic acid has a carboxylate group on the side chain. Thus this tripeptide has three carboxylate groups and one amino group that are all charged at physiological pH. The tripeptide has a charge of 2– and will migrate to the positive anode.

$$H_3N^+-CH-\overset{\overset{O}{\|}}{C}-\overset{\overset{H}{|}}{N}-CH-\overset{\overset{O}{\|}}{C}-NH-CH-\overset{\overset{O}{\|}}{C}-O^-$$
$$\overset{|}{CH_2COO^-} \quad \overset{|}{CH_3} \quad \overset{|}{CH_2COO^-}$$

20.21 In digestion, acidic hydrolysis, and basic hydrolysis, the peptide bonds are broken to leave the individual amino acids Phe, Asp, Thr, and Lys. However, in acidic hydrolysis, the acidic environment results in the amino groups of the resultant amino acids being charged, whereas the carboxylate groups are uncharged. In basic hydrolysis, the basic environment results in the carboxylate groups of the resultant amino acids being charged, whereas the amino groups are uncharged.

20.23 Hydrolysis breaks the peptide bonds to show the composition of the original tripeptide; it must consist of Ala, Gly, and Lys. Hydrolysis, however, does not show the order in which these three amino acids were bonded together; that is, the sequence. Thus the original tripeptide may be Ala-Gly-Lys, Ala-Lys-Gly, Lys-Ala-Gly, Lys-Gly-Ala, Gly-Ala-Lys, or Gly-Lys-Ala.

20.25 (a) Digestion breaks the peptide bonds to leave:

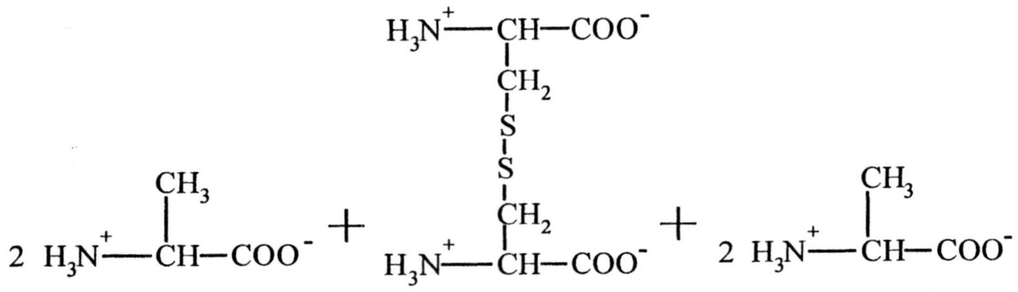

$$2\ H_3\overset{+}{N}-\underset{\underset{CH_3}{|}}{CH}-COO^- \quad + \quad H_3\overset{+}{N}-\underset{\underset{\underset{\underset{\underset{\underset{H_3\overset{+}{N}-CH-COO^-}{|}}{CH_2}}{|}}{S}}{|}}{CH}-COO^- \quad + \quad 2\ H_3\overset{+}{N}-\underset{\underset{CH_3}{|}}{CH}-COO^-$$

(b) There are no functional groups in this hexapeptide that will selectively oxidize; thus there is no reaction.

(c) The disulfide bond can be reduced to form two tripeptides:

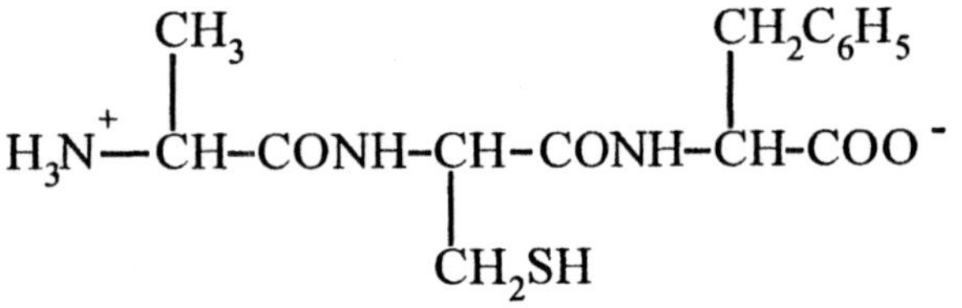

$$H_3\overset{+}{N}-\underset{\underset{CH_3}{|}}{CH}-CONH-\underset{\underset{CH_2SH}{|}}{CH}-CONH-\underset{\underset{CH_2C_6H_5}{|}}{CH}-COO^-$$

20.27 Simple proteins contain only molecules that are made up of amino acids, whereas conjugated proteins contain nonpeptide molecules or ions as well as peptide molecules.

20.29 Conformation defines the three-dimensional structure of a large molecule. This three-dimensional structure can be changed by rotation about a single bond.

20.31 Hydrogen bonding between peptide groups (a) is most important in determining the secondary structure of polypeptides.

20.33 (a) Proline is nonpolar, whereas histidine is a basic amino acid; thus there is no attractive interaction between these two amino acids.

(b) Serine has an –OH group and tyrosine has an –OH; thus hydrogen bonds can form between these two amino acids.

(c) Both proline and phenylalanine have nonpolar side groups; thus hydrophobic interactions can take place between these two amino acids.

(d) Both lysine and arginine are basic amino acids; thus there is no attractive interaction between these two amino acids.

(e) Lysine is a basic amino acid, whereas glutamic acid is an acidic amino acid; thus

Proteins 161

these two amino acids can undergo an acid-base reaction to form a salt bridge between the two amino acids.

(f) Serine is a polar amino acid, whereas valine is a nonpolar amino acid; thus there is no attractive interaction between these two amino acids.

20.35 Nonpolar residues must be kept from water, hydrogen bonds can form between peptide units, and polar residues can interact with water.

20.37 The primary structure defines the secondary, tertiary, and quaternary structure of a protein.

20.39 Fibrous proteins can aggregate by strong secondary attractive interactions to form macroscopic structures that are strong. This formation of strong macroscopic structures is imperative in the function of structural and contractile proteins.

20.41 A double helix is composed of two polypeptide chains that are coiled around each other, whereas a triple helix is composed of three polypeptide chains that are coiled around each other.

20.43 α-Keratins provide structural shape to cilia, hair, horns, hooves, skin, and wool.

20.45 When proteins function as catalysts, regulators, transporters, or protectors, they must not aggregate together to form macroscopic structures. The globular shape of globular proteins is not conducive to aggregation and thus they can fulfill this requirement. Their globular structure is also conducive to dissolution owing to their hydrophilic surfaces.

20.47 Tertiary structure describes the relation between different conformational patterns within a given polypeptide. α-Keratins form a helix throughout the polypeptide chain and therefore do not possess different conformational patterns. Myoglobin, however, contains helical parts and β-sheet parts in different parts of the polypeptide chains. The arrangement of these helical and β-sheet structures in space is the tertiary structure of myoglobin.

20.49 The prosthetic group is the nonpeptide part of a conjugated protein, whereas the polypeptide is the apoprotein. Thus, the polypeptide is the apoprotein, and the heme is the prosthetic group.

20.51 Heme has an Fe^{2+} ion that bonds with the oxygen to hold it.

20.53 The amino acid residues that are on the exterior of globular proteins must be polar groups so that they can interact preferentially with water to dissolve the protein. Thus, Asp, Glu, His, and Lys are on the exterior of the globule.

20.55 A mutation is an alteration in the DNA structure of a gene. This alteration sets off a chain of events; it will produce a change in the primary structure of the protein encoded by the DNA, and this change may in turn alter the function of that protein.

20.57 Serine is a polar residue, whereas both phenylaniline and isoleucine are nonpolar residues. Thus the replacement of Phe with Ser alters the polar properties of that part of the peptide chain. However, if Ile replaces Phe, that part of the peptide still remains nonpolar.

20.59 The part of the peptide where this mutation occurs will become more polar. The peptide will thus attempt to move this part to the exterior of the protein, which results in a change in the three-dimensional structure of the protein. This in turn will alter the its function.

20.61 A native protein is one that has the conformation that exists under physiological conditions.

20.63 Digestion breaks the peptide linkages in a protein to alter its primary structure and produce amino acids. Denaturation alters the interactions between residues that give a protein its secondary, tertiary, and quaternary structures. Thus denaturation changes the secondary, tertiary, and quaternary structures.

20.65 Ag^+ reacts with sulfhydryl groups of cysteine residues within the gonorrhea organism to form metal disulfide bridges. These metal disulfide bridges alter the conformation of a protein, which in turn alters its function. This results in the death of the gonorrhea organism.

20.67 A change in pH will alter the charged state of the basic and acidic groups of the amino acid residues. Thus, a change in pH will alter the salt bridges that are formed from the charged acidic and basic groups in the residues. If a peptide does not have many acidic and basic residues, pH will not alter the structure.

20.69 To denote D and L designation, the structure must be drawn such that –COOH and –CH$_2$CH$_2$COOH are vertical in the structure. In the L structure, NH$_2$ is on the left. Thus, the structure is:

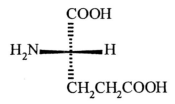

Proteins 163

20.71 (a) At its isoelectric point, methionine exists as a zwitterion. At pH < 1, the solution is very acidic. Under these conditions, the negatively charged carboxyl group reacts with free protons, which results in an amino acid that has a net positive charge. At pH > 12, the solution is very basic, and the positively charged amino group will give up a proton to leave an amino acid that that has a net negative charge. Thus:

$$\underset{pH < 1}{H_3\overset{+}{N}-\underset{|}{\overset{CH_2CH_2SCH_3}{CH}}-COOH} \qquad \underset{pH \sim pI \text{ and } 7}{H_3\overset{+}{N}-\underset{|}{\overset{CH_2CH_2SCH_3}{CH}}-COO^-} \qquad \underset{pH > 12}{H_2N-\underset{|}{\overset{CH_2CH_2SCH_3}{CH}}-COO^-}$$

(b) Methionine is a neutral compound at pI but is an ion at higher and lower pH. Ions are more soluble in water than are neutral compounds; thus methionine is least soluble in water at pI. It will be most soluble at very high and very low pH.

(c) At pI, the molecule is neutral and does not migrate in electrophoresis. At pH = 7, the pH is within 2 of pI, and thus the molecule will not migrate in electrophoresis. At pH = 1, methionine is positively charged and thus will migrate to the negatively charged cathode. At pH = 12, the amino acid is negatively charged and will migrate to the positively charged anode under this condition.

20.73 There are 5! different constitutional isomers = 120. The two isomers that have Val as the N-terminal, proline as the C-terminal, and lysine as the middle residue are Val-Phe-Lys-Thr-Pro and Val-Thr-Lys-Phe-Pro.

20.75 Glycine has two –H atoms bonded to the carbon and thus does not have a tetrahedral stereocenter. Therefore, in this tripeptide, only Lys and Ala can attain either L or D configuration. The four stereoisomers are thus L-Lys-Gly-L-Ala, D-Lys-Gly-D-Ala, D-Lys-Gly-L-Ala, and L-Lys-Gly-D-Ala. Only the isomer with all L configuration amino acids will be found in nature, L-Lys-Gly-L-Ala.

20.77 • At pH = 7, the carboxylate and amino groups are charged.

• To name peptides, (i) the C-terminal residue keeps its amino acid name, (ii) the -ine or -ic acid of the other amino acid residues is replaced with -yl, and (iii) naming begins at the N-terminal residue.

Thus, Met-Ser-Ile-Gly-Glu is methionylserylisoleucylglutamic acid and has a structure of:

$$H_3N^+-\underset{\underset{CH_2CH_2SCH_3}{|}}{CH}-CONH-\underset{\underset{CH_2OH}{|}}{CH}-CONH-\underset{\underset{CHCH_3}{|}}{\underset{|}{\overset{CH_2CH_3}{|}}}CH-CONH-CH_2-CONH-\underset{\underset{CH_2CH_2COO^-}{|}}{CH}-COO^-$$

20.79 At pH = 7, the carboxylate and amino groups are charged. Therefore its structure is:

$$NH_3^+-\underset{\underset{(CH_2)_4NH_3^+}{|}}{CH}-\overset{O}{\overset{||}{C}}-\underset{\underset{}{|}}{\overset{H}{\overset{|}{N}}}-\underset{\underset{CH_3}{|}}{CH}-\overset{O}{\overset{||}{C}}-NH-\underset{\underset{(CH_2)_4NH_3^+}{|}}{CH}-\overset{O}{\overset{||}{C}}-O^-$$

There is one more amino group than carboxylate; thus the peptide has an overall charge of 1+ and migrates to the negatively charged cathode. The pI of this compound is slightly more basic than that of a neutral background to account for the 1+ charge.

20.81 Phe-Asp-Leu-Glu-Thr has two more carboxylate groups (from Glu and Asp) than amino groups and will therefore have a 2– charge.

20.83 When this peptide is reduced, the disulfide bridge will be broken to form the following peptide:

Gly-Ala-Cys-Gly-Lys-Phe-His-Glu-His-Cys-Met-Gly-Asp

20.85 (a) Albumin is a neutral compound at pI but is an ion at higher and lower pH. Ions are more soluble in water than neutral compounds; thus albumin is least soluble in water at pI = 4.9.

(b) At physiological pH, the compound is more basic than at its pI and is therefore negatively charged. Therefore, it will migrate to the positively charged anode.

20.87 The compound must be negatively charged at physiological pH to migrate to the positively charged anode. Therefore, the compound must be in a more basic environment than its pI. Therefore, the compound must be pepsin.

20.89 Globular proteins require water solubility, whereas fibrous proteins do not; thus globular proteins must have more polar residues on their surfaces.

20.91 X cannot be a β-pleated sheet, because it has a low content of Gly, Ala, and Ser. It cannot be an α-helix either, because it contains significant amounts of the helix breakers proline and glutamic acid. Therefore X is either a β-turn or a loop.

20.93 (a) Trp is a very large residue, whereas Gly is a very small residue. This replacement of a large residue with a small residue will alter the three-dimensional structure of the protein and thus its function.

(b) Both Arg and Lys are similarly sized basic residues, and thus their exchange does not alter the structure of the protein.

(c) Both Thr and Ser are similarly sized neutral polar residues, and thus their exchange does not alter the three-dimensional structure and function of the protein.

(d) Arg is a basic residue, whereas Asp is an acidic residue. This exchange for acid and base will alter salt-bridge interactions that control the three-dimensional structure, which in turn will alter the function of the protein.

20.95 Residue 1 must be formed from lysine by addition of an –OH group to the first carbon in the side group. Residue 2 must be formed from proline by addition of an –OH group to the cyclopentane ring. Residue 3 must be formed from glutamic acid by addition of another acidic group to the side chain.

20.97 (a) The interior of the bilayer is hydrophobic and thus must consist of nonpolar residues such as Leu, Phe, and Val.

(b) The extracellular environment is aqueous and thus must consist of hydrophilic residues such as Glu, Lys, and Ser.

(c) The intracellular environment is aqueous and thus must consist of hydrophilic residues such as Glu, Lys, and Ser.

20.99 The alteration in sickle-cell anemia hemoglobin is in a region that strongly influences the conformation and function of the protein. The difference in residues between beef and human insulin must be in regions that do not significantly alter the conformation and function of the protein.

20.101 Proteins are formed from many residues and are much larger molecules than the lipids.

Chapter 21

Nucleic Acids

21.1 Nucleotides are composed of a five-carbon sugar, phosphoric acid, and a heterocyclic nitrogen base. The five-carbon sugar in ribonucleotides is ribose, whereas the five-carbon sugar in deoxyribonucleotides is deoxyribose.

21.3 In a nucleotide, the phosphoric acid, sugar, and base are bonded together. Thus, the nucleotide that is composed of phosphoric acid, deoxyribose, and thymine is:

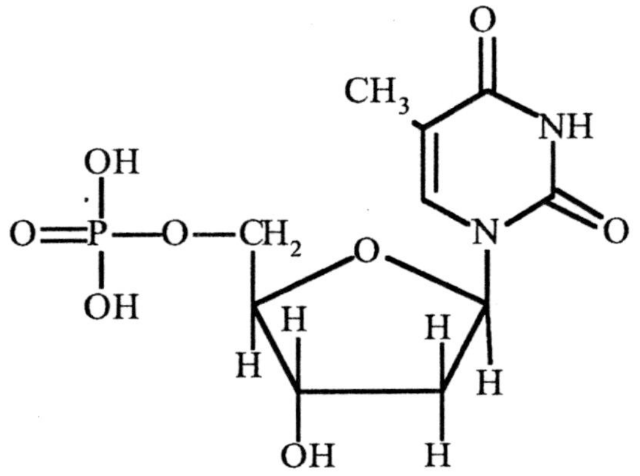

The abbreviated name consists of a "d" to denote deoxyribose followed by the one letter abbreviation for the base followed by "MP," which denotes 5'-monophosphate. Thus, the abbreviated name is dTMP.

21.5 dGMP is composed of guanine (G), deoxyribose (d), and phosphoric acid (MP). Hydrolysis breaks the bonds that were created in the formation of the nucleotide and thus results in the products: deoxyribose, phosphoric acid, and guanine.

21.7 DNA is found only in the nucleus of the cell because it is too big to escape.

21.9 This compound is two nucleotides bonded together. The upper nucleotide consists of phosphoric acid, deoxyribose, and thymine (dTMP), and the lower nucleotide consists of phosphoric acid, deoxyribose, and cytosine (dCMP). Therefore, this compound is dTMP-dCMP.

21.11 dUMP-dGMP contains deoxyribose and therefore would be a part of DNA. However, DNA does not contain uracil (U); therefore, this sequence cannot exist in nature.

21.13 This diad consists of a nucleotide, in the 5' position, that contains guanine and a nucleotide that contains thymine.

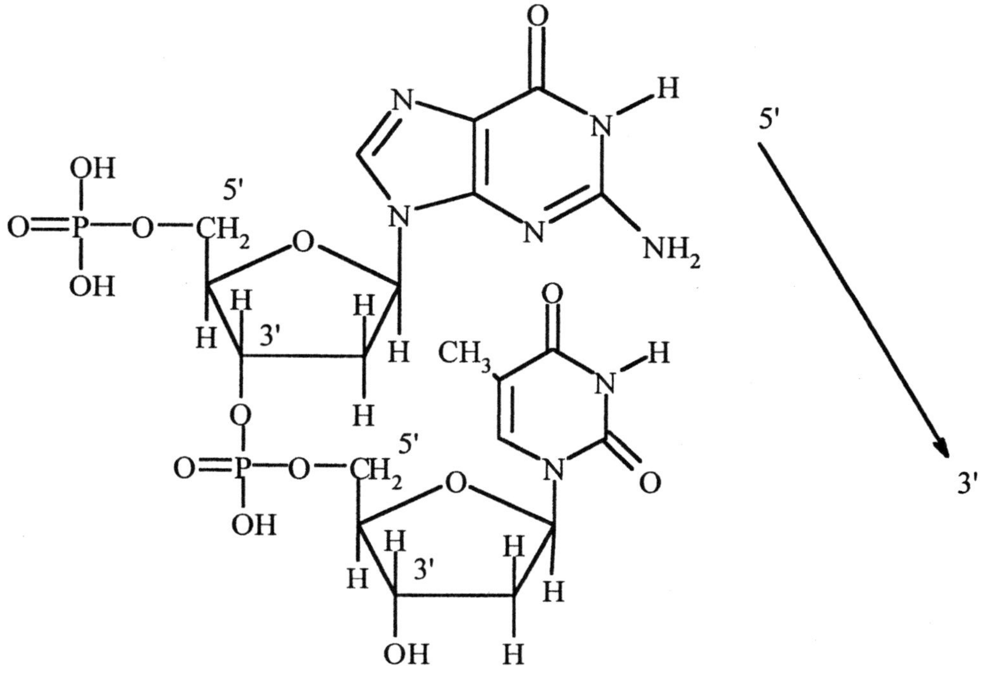

[Structural diagram of a dinucleotide: 5'-guanine nucleotide linked via phosphodiester bond to 3'-thymine nucleotide, with an arrow indicating 5' to 3' direction]

21.15 Hydrolysis breaks the linkage between nucleotides; thus it produces dAMP, dGMP, dCMP, and two dTMP molecules.

21.17 Histones are proteins. Their function is to compress the large DNA double helix molecule into the small volume of the cell nucleus.

21.19 Base-pairing is the result of intermolecular hydrogen bonding.

21.21 d. It cannot be part a, because A and T are always present in 1:1 ratio; it cannot be part b, because the ratio of sugars to bases is 1:1; it cannot be part c, because, in a given nucleic acid, all of the sugars are the same.

21.23 Adenine and thymine exist in a 1:1 ratio, and cytosine and guanine also exist in a 1:1 ratio. Therefore, there must be 30.4 mol-% thymine and 19.6 mol-% cytosine.

21.25 The complementary base pair for A is T, for G is C, for C is G, and for T is A. Therefore, the complementary strand is 5'-CGGTTA-3'.

21.27 Genetic information is stored in the sequence of bases along the DNA chain.

21.29 Replication produces a chromosome.

168   Chapter 21

21.31 (a) A replication bubble is the part of the DNA double helix that is unwound and is where replication takes place.

(b) A leading strand is the DNA strand that is synthesized in a continuous manner; a lagging strand is the strand that is synthesized in a discontinuous manner.

(c) Okazaki fragments are the discontinuous DNA fragments of the lagging strand.

21.33 Parent DNA consists of two DNA strands that form a double helix. In replication, this parent separates two DNA strands that pair with two new DNA strands to create a pair of daughter DNA molecules.

21.35 (a) The corresponding sequence in the old DNA chain of the daughter double helix consists of complimentary base pairs and thus is 5'-CGCTAA-3'.

(b) The corresponding sequence in the old DNA chain of the other daughter double helix must consist of the same sequence and thus is 5'-TTAGCG-3'.

21.37 Transcription produces RNA from DNA.

21.39 The template DNA strand is used to produce the primary transcript.

21.41 Posttranscriptional processing modifies various primary transcripts to form rRNA, mRNA, and tRNA.

21.43 DNA is orders of magnitude larger than any RNA; DNA typically has a molecular mass of many billions, rRNA typically has a molecular mass of 500,000 to 1 million; mRNA typically has a molecular mass between 100,000 and 1or 2 million; tRNA typically has a molecular mass less than 50,000.

21.45 In RNA, U forms a base pair with A, and C forms a base pair with G. Therefore, the composition of RNA formed from this sequence is 15 mol-% U, 25 mol-% G, 20 mol-% C, and 40 mol-% A.

21.47 Translation is the synthesis of polypeptides.

21.49 rRNA and proteins form ribosomes, in which polypeptide synthesis takes place. mRNA carries the genetic message that encodes the amino acid sequence of the polypeptide to be synthesized. mRNA also binds to the ribosome and acts as the template for polypeptide synthesis. tRNA transports α-amino acids into the ribosome.

21.51 The sequence of bases in mRNA is the genetic message that it carries. This information specifies the sequence of α-amino acids in the polypeptide to be synthesized.

Nucleic Acids 169

21.53 The genetic code is considered universal, because almost every living species, including plants and animals, follow the same genetic code.

21.55 Aminoacyl-tRNA synthetase is the enzyme that catalyzes the concatenation of amino acids at aminoacyl and peptidyl sites on mRNA.

21.57 A base triplet is the name given to three consecutive bases along a DNA or RNA chain.

21.59 The primary transcript is merely a chain with base pairs that are complementary in RNA to the template DNA strand; thus the primary transcript is 3'-UGU-GUG-GUU-UAC-ACA-CCA-GUA-5'.

The mRNA is then formed only from the exons of the template DNA strand. It is formed from the complementary base pairs and is therefore 5'-AUG-CAU-UUG-UGU-3'.

This mRNA sequence encodes the polypeptide Met-His-Leu-Cys.

21.61 A mutation is an error in the base sequence of a gene.

21.63 In a substitution mutation, the identity of a base along a sequence is changed. In a frameshift mutation, a base is inserted or deleted into a sequence such that the sequence is shifted.

21.65 A spontaneous mutation is one that results from a random error in replication.

21.67 Germ cells are egg and sperm cells that contain information that can be passed onto offspring or future generations. Somatic cells are all other cells.

21.69 A cancer is an uncontrolled growth and division of cells to form tumors.

21.71 Biological function is most often affected by a mutation when a residue is replaced by another that is of different size, charge, or polarity. In mutation a, a negatively charged amino acid residue is replaced by an uncharged residue; whereas, in mutation b, one negatively charged residue is replaced by another. Thus, mutation a is more harmful.

21.73 The peptide that is synthesized is determined by the sequence of base pairs in each codon that results from the exons. This procedure begins at the 3' end. Thus, the first exon (z) is TAC, which correlates with the codon AUG (T pairs with A, A pairs with U, and C pairs with G). Exons Z and Y (CTG and AAT) results in the codons GAC and UUA.

(a) With reference to Table 21.1, AUG encodes the synthesis of Met, GAC encodes Asp, and UUA encodes Leu. Thus, Met-Asp-Leu will be synthesized from this DNA strand.

(b) If the GTC triplet is mutated to CTC, this exon will produce the codon GAG, which will result in the synthesis of Glu rather than Asp; thus the polypeptide Met-Glu-Leu will be synthesized.

(c) If GTC triplet is mutated to ATC, this exon will produce the codon GAU, which will result in the synthesis of Asp; thus the polypeptide Met-Asp-Leu will still be synthesized.

(d) There is no effect of a mutation in the intron on the process of exon → codon → polypeptide; thus the polypeptide Met-Asp-Leu will still be synthesized.

21.75 An antibiotic is a chemical, usually an organic compound, that kills microorganisms such as bacterium, mold, or yeast.

21.77 A virus is an infectious, parasitic particle consisting of either DNA or RNA, but not both, and encapsulated in a protein coat. A virus is also usually smaller than bacteria.

21.79 A vaccine contains a weakened virus or its proteins. Its purpose is to prevent viral infections by stimulating the immune system to generate antibodies that will recognize the virus.

21.81 Recombinant DNA technology relies on the principle that transplanting DNA from one organism into another can alter the genome of the other organism.

21.83 In recombinant DNA technology, the vector DNA is the organism's DNA that is to be altered. The donor DNA is the DNA that is introduced into the vector DNA to create this alteration. The altered DNA is the recombinant DNA.

21.85 The bonding of many nucleotides together forms nucleic acids. Thus, nucleotides are building blocks for nucleic acids.

21.87 At the 5'-end of a nucleic acid only one of the oxygens of the phosphate group at C-5 is attached to a carbon; at the 3'-end of the nucleic acid, an unreacted –OH is attached to C-3 of the sugar.

21.89 This abbreviated sequence contains deoxyribose and is thus part of DNA. However, DNA is formed from C, T, A, and G. There are no deoxyribonucleic acids that contain U.

21.91 Hydrolysis breaks the linkage between nucleotides; thus it will produce dAMP, dCMP, dGMP, and dTMP in a ratio of 1:1:2:3.

21.93 DNA exists in the form of a double helix consisting of two DNA strands, whereas RNA is single stranded.

21.95 Adenine forms a base pair with thymine, whereas cytosine forms a base pair with guanine. Thus, for every adenine in one strand, there is a thymine in the other strand. The same relation holds for cytosine and guanine between the two strands.

(a) 14 mol-% A and 31 mol-% C

(b) 17 mol-% C and 38 mol-% A

21.97 The primary transcript is an RNA chain with bases that are complementary to the template DNA strand; thus the primary transcript is 3'-AUG-ACC-ACA-CAA-UUU-GUG-AGU-5'.

The mRNA is then formed only from the exons of the template DNA strand. It is formed from the complementary bases and is therefore 5'-AUG-CAA-UUU-AGU-3'.

This mRNA sequence encodes the polypeptide, Met-Glu-Phe-Ser.

21.99 A transgenic plant or animal conceived from a germ cell containing recombinant DNA.

21.101 Gene therapy seeks to correct hereditary diseases by introducing recombinant DNA into an organism.

21.103 Denaturation occurs when the hydrogen bonds that stabilize base pairs are broken. G-C base pairs are stabilized by three hydrogen bonds, whereas A-T base pairs have two. This means that it is easier to break A-T pairs than G-C pairs. Therefore, the more G-C pairs that exist, the more thermal energy (and higher temperature) is needed to break this stronger interaction.

21.105 DNA consists of pairs of DNA molecules held together by the specific base pairs G-C and A-T. This specific pairing results in the number of G being equal to the number of C and the number of A being equal to the number of T in a given DNA double helix. RNA, however, exists as single molecules and thus does not require these specific ratios.

21.107 The synthesis of each amino acid residue is encoded by a three-nucleotide-base sequence; thus 146 amino acids will require at least 438 (= 3 × 146) nucleotide bases. There are usually many more nucleotide bases because of the introns present. Only the exons of the gene encode the synthesis of the polypeptide.

Chapter 22

Metabolism and Enzymes: An Overview

22.1 Biochemistry is the extension of the relation of the structure and function of cells to their organization on molecular and subcellular levels.

22.3 Eukaryotic cells contain subcellular membrane-bounded organelles or structures, whereas prokaryotic cells do not.

22.5 Oxidative processes take place within the mitochondria of animal cells.

22.7 Protein synthesis takes place at ribosomes.

22.9 DNA in bacterial cells is not bounded by a membrane and is located in a microscopically visible nuclear zone.

22.11 (i) To obtain energy in a chemical form by the degradation of nutrients.

(ii) To convert nutrient molecules into precursor molecules used to build cell macromolecules.

(iii) To synthesize cell macromolecules.

(iv) To produce or modify the biomolecules necessary for specific functions in specialized cells.

22.13 Anabolism is the group of processes that synthesize biomolecules.

22.15 The first stage of catabolism reduces nutrient macromolecules to monosaccharides, amino acids, and fatty acids.

22.17 Acetyl-S-CoA is oxidized to carbon dioxide (CO_2), and water (H_2O).

22.19 ATP functions as the carrier of energy to the energy-requiring processes of cells and is the link between catabolism and anabolism.

22.21 ATP can be synthesized by substrate phosphorylation, in which a phosphate group is added to ADP from a high-energy compound, and by oxidative phosphorylation, in which ADP reacts directly with an inorganic phosphate.

22.23 NAD^+ functions in catabolic reactions, whereas the reduced form of $NADP^+$ (NADPH) is primarily used in reductive biosynthetic reactions.

22.25 An active site is a region on an enzyme where catalysis takes place. It is a binding site and it has catalytic function.

22.27 Inducible fit describes cases in which the structure of a substrate molecule can influence the complementarity of a binding site.

22.29 Many vitamins become cofactors or serve as sources that can be transformed into cofactors.

22.31 The turnover number is the number of moles of substrate that react per unit time per mole of enzyme.

22.33 A regulatory enzyme is one whose activity can be modified by combination with specific activators or inhibitors.

22.35 High-energy compounds are responsible for driving forward chemical reactions that are unfavorable.

22.37 Cells handle the different types of molecules by breaking down the incoming compounds and dealing with the products in similar ways. Catabolism first collects functional molecules by degrading nutrient macromolecules to their building blocks: proteins to amino acids, and polysaccharides to monosaccharides, and hydrolyzable lipids to fatty acids and glycerol. After that, hexoses, the carbon chains of fatty acids, and most of the amino acids are converted into acetyl-S-CoA, which is eventually oxidized to carbon dioxide and water.

22.39 No. The activities of cellular enzymes are regulated so that they can respond appropriately to the immediate metabolic needs of the cell. This regulation can be by various methods, including the binding of molecules that are not their substrates to allosteric regulatory enzymes, covalent modification, or feedback inhibition.

Chapter 23

Carbohydrate Metabolism

23.1   Glucose is the principal source of energy for metabolism.

23.3   The most important products of glycolysis are ATP, NADH, and lactate (pyruvate under aerobic conditions).

23.5   The activity of the enzyme phosphofructokinase decreases at high concentrations of ATP. Therefore, this enzyme is the key regulatory control point in glycolysis.

23.7   In glycolysis, ATP is synthesized in two reactions, both of which take place in the second stage of glycolysis:

ADP + 3-bisphosphoglycerate → ATP + 3-phosphoglycerate

ADP + phosphoenolpyruvate → ATP + pyruvate

23.9   In active skeletal muscles, NAD^+ is regenerated by the reduction of pyruvate to lactate.

NADH + H^+ + pyruvate → NAD^+ + lactate

23.11 Glycolysis would cease if inorganic phosphate were unavailable.

23.13 The reaction of pyruvate with NAD^+ and CoA-SH to form acetyl-S-CoA takes place in mitochondria.

23.15 Yes. It is deactivated by phosphorylation and activated by dephosphorylation.

23.17 Carbon dioxide is formed in two reactions of the citric acid cycle, when isocitrate reacts with NAD^+ to form α-ketoglutarate:

Isocitrate + NAD^+ → α-ketoglutarate + CO_2 + NADH

and when α-ketoglutarate reacts with CoA-SH to form succinyl-S-CoA:

α-Ketoglutarate + CoA-SH → succinyl-S-CoA + CO_2

23.19 Because the products of the sequence of reactions form a circle. In other words, the product of the sequence, oxaloacetate, is also the first reactant of the same sequence.

23.21 Because oxaloacetate is regenerated at the conclusion of each cycle, one molecule of it can effect the oxidation of a limitless numbers of acetyl groups. Therefore, oxaloacetate acts a catalyst for this cycle.

Carbohydrate Metabolism

23.23 The concentration of oxaloacetate increases when the enzyme pyruvate carboxylase is activated by acetyl-S-CoA and, in the presence of ATP, catalyzes the formation of oxaloacetate from pyruvate and CO_2.

23.25 Not directly. Oxaloacetate is first synthesized in mitochondria and reduced to malate, which then enters the cytosol and is reoxidized to oxaloacetate. There the oxaloacetate reacts with GTP under the influence of the enzyme phosphoenolpyruvate carboxykinase to form phosphoenolpyruvate, CO_2, and GDP.

23.27 Oxaloacetate is first reduced to malate, which can pass through the mitochondrial membrane and enter the cytosol.

23.29 Glycogen is the polymeric storage form of glucose in animal tissue.

23.31 The hydrolysis of pyrophosphate has a very large equilibrium constant, is coupled to the overall metabolic process, and is the driving force for the formation of the activated glucose molecule.

23.33 In phosphorolysis, a glucose molecule is removed from glycogen by the formation of glucose-1-PO_4, whereas hydrolysis yields glucose.

23.35 Glucagon and epinephrine are hormones that initiate glycogenolysis.

23.37 They both regulate the blood-glucose concentration. Epinephrine also affects blood pressure and heart rate.

23.39 Only liver cells possess the enzyme glucose-6-phosphate phosphatase, which hydrolyzes glucose-6-PO_4 to form free glucose.

23.41 There are three locations along the electron-transport chain where there is sufficient energy for the synthesis of ATP.

23.43 No. The inner membrane of a mitochondrion must be intact for ATP synthesis to take place.

23.45 No. The inner membrane is impermeable to most compounds. Transport across this membrane is through transport proteins that are specific for only a few substances.

23.47 The maximum total amount of ATP produced under aerobic conditions is 38 mol of ATP per mole of glucose.

23.49 In active muscle cells, the rate of glycolysis is much greater than the rate of the citric acid cycle.

23.51 Active muscle cells produce ATP primarily through glycolysis. For glycolysis to continue at a maximal rate, NAD^+ must be regenerated by oxidation of NADH and inorganic phosphate must be available for the formation of 1,3-bisphosphoglycerate. Therefore, for muscle cells to produce ATP, NAD^+ must be regenerated quickly and inorganic phosphate must be available.

23.53 NADH delivers its reducing power to the electron-transport chain by reducing oxidized cytosolic substances to their reduced counterparts, which can then penetrate the mitochondrial membrane.

Chapter 24

Fatty Acid Metabolism

24.1 No. The upper limit for the mass of glycogen stored in the liver and muscles is only about 12 hours.

24.3 When the blood-glucose concentration reaches its lowest levels between meals, glucagon is released.

24.5 A fatty acid is oxidized only in the form of an acyl-S-CoA thioester; the reaction that activates the fatty acid to oxidation is:

R-COO⁻ + ATP + CoA-SH → RCO-S-CoA + AMP + 2 P_i

24.7 No. There is no specific transport system for fatty acyl-S-CoA derivatives in the mitochondrial membrane.

24.9 The first step in fatty acid oxidation is a dehydrogenation between carbons 2 and 3. Thus, the product for a 16-carbon fatty acid is:

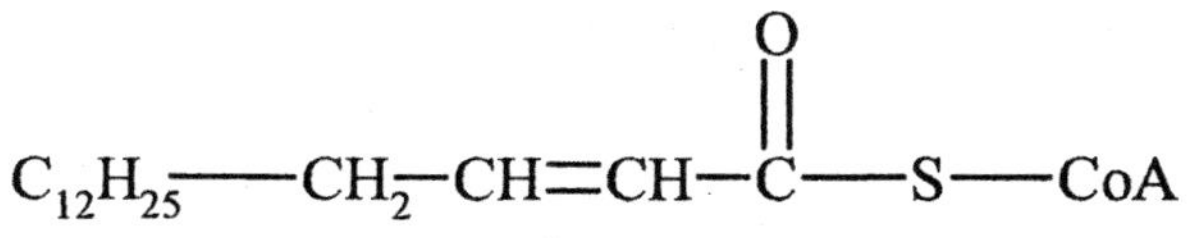

$$C_{12}H_{25}-CH_2-CH=CH-\overset{\overset{O}{\|}}{C}-S-CoA$$

trans-Δ^2-Enol-CoA

24.11 The second step is the enzymatic hydrogenation of the trans double bond to form:

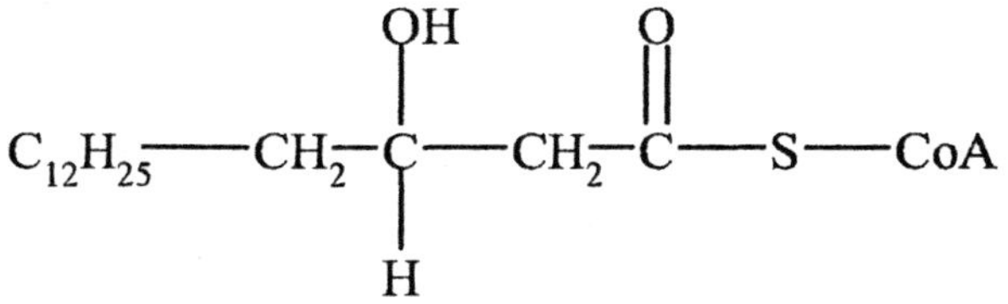

$$C_{12}H_{25}-CH_2-\overset{\overset{OH}{|}}{\underset{\underset{H}{|}}{C}}-CH_2-\overset{\overset{O}{\|}}{C}-S-CoA$$

L-3-Hydroxyacyl-S-CoA

24.13 The third step is a dehydrogenation to form:

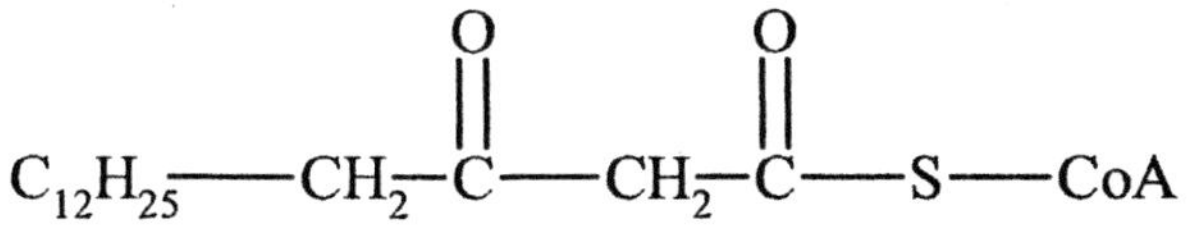

$$C_{12}H_{25}-CH_2-\overset{\overset{O}{\|}}{C}-CH_2-\overset{\overset{O}{\|}}{C}-S-CoA$$

3-Ketoacyl-S-CoA

24.15 The reactants in the fourth oxidation step are the product of the third step, 3-keto(C_{16})acyl-S-CoA, and CoA-SH.

24.17 The ketone bodies synthesized in the liver are acetoacetate, D-3-hydroxybutyrate, and acetone.

24.19 Ketosis occurs when acetyl-S-CoA is in excess, which usually occurs during starvation and in diabetes mellitus.

24.21 Fatty acid synthesis takes place in the cytosol, whereas oxidation or catabolism takes place in mitochondria.

24.23 Insulin initiates fatty acid biosynthesis by activating the phosphodiesterase.

24.25 Citrate from mitochondria serves as the principal source of acetate in fatty acid biosynthesis.

24.27 Acetyl-S-CoA is transferred to the α-SH site, and malonyl-S-CoA is transferred to the β-SH site.

24.29 CO_2 is lost as HCO_3^-, which makes the reaction irreversible and provides the driving force for the reaction. Biotin is the cofactor required for the formation of malonyl-S-CoA.

24.31 Prior to chain lengthening, the α-SH site is occupied by an acetyl group, and a carboxylated acyl derivative is at the β-site.

24.33 The first step in the biosynthesis of triacylglycerols begins with the formation of phosphatidate from glycerol-3-PO_4 and acyl-S-CoA.

24.35 Yes, it is worth the investment; the synthesis of a triacylglycerol requires about 15% of the ATP that is generated in its oxidation.

24.37 Ethanolamine is added to diacylglycerol in the activated form of cytidine diphosphoethanolamine.

24.39 The hydrolysis of complex cellular glycolipids that normally takes place in the lysosomes does not take place.

24.41 In catabolism, the acyl carrier is CoA-SH, but, in biosynthetic reactions, the acyl carrier is an –SH protein. Reduction in catabolism employs NADH, whereas NADPH is used in biosynthetic reactions.

24.43 8 Acetyl-S-CoA + 7 ATP^{4-} + 14 NADPH + 7 H^+ → palmitoyl-S-CoA + 14 $NADP^+$ + 7 CoA-SH + 7 ADP^{3-} + 7 P_i^{2-}

24.45 Yes. The oxidation of the fatty acid chain produces NADH and $FADH_2$, both of which provide reducing power to the electron-transport chain for the synthesis of ATP through oxidative phosphorylation.

24.47 True. The presence of high concentrations of citrate indicates that the cells are in a high-energy state (ATP in high concentration). Equally important is the fact that citrate is a specific allosteric activator of acetyl-S-CoA carboxylase, which catalyzes the rate-limiting step in the fatty acid synthase system.

24.49 Fatty acids arise in adipose tissue by enzymatic hydrolysis of the stored triacylglycerols. The hydrolysis is catalyzed by a lipase that is activated by glucagon. The fatty acids leave adipose cells, become solubilized by being bound to serum albumin, and in that form travel throughout the circulatory system.

Chapter 25

Amino Acid Metabolism

25.1 Transamination is a process in which the amino group of an amino acid is interchanged with the carbonyl group of an α-keto acid.

25.3 Excess amino acids not used in biosynthesis are catabolized and used as energy sources.

25.5 The first step in the catabolism of ingested amino acids is the transfer of the amino groups to α-ketoglutarate by transamination.

25.7 NH_4^+ from liver cells is rendered nontoxic by conversion into urea.

25.9 A urotelic animal excretes ammonium ion in the form of urea.

25.11 If there is an excess concentration of NH_4^+ in cells, it reacts with a large amount of α-ketoglutarate. This results in the excessive lowering of α-ketoglutarate concentrations.

25.13 Muscle cells use the glucose-alanine cycle to render NH_4^+ harmless for transport through the blood.

25.15 The reactions constituting the urea cycle begin in the cytosol, continue in mitochondria, and end in the cytosol.

25.17 Citrulline is an amino acid that participates in urea's synthesis

$$H_3N^+-\underset{H}{\overset{COO^-}{\underset{|}{C}}}-CH_2-CH_2-CH_2-\underset{}{\overset{H}{N}}-\overset{O}{\underset{}{C}}-NH_2$$

25.19 ATP is not required for the formation of citrulline, the formation of arginine and fumarate from argininosuccinate, and the hydrolysis of arginine to form urea and ornithine.

25.21 Approximately 15% of the energy available in amino acids is used to synthesize urea.

25.23 The catabolism of these amino acids provides products that can be used to synthesize glucose.

25.25 The catabolism of these amino acids results in the formation of ketone bodies.

25.27 When phenylalanine is catabolized, acetoacetyl-S-CoA and fumarate are formed.

Fumarate can be used to create glucose through gluconeogenesis, and acetoacetyl-S-CoA yields ketone bodies in the liver. Therefore the products of phenylalanine's catabolism result in the formation of both glucose and ketone bodies.

25.29 The metabolic defect causes the intermediates of catabolic or anabolic sequences to accumulate in cells.

25.31 Phenylketonuria leads to severe mental retardation. Treatment requires that phenylalanine be eliminated or severely restricted from the diet of newborns.

25.33 A nonessential amino acid is one that can be synthesized by humans.

25.35 The nonionic water-soluble compound urea can be excreted without the loss of other important anions such as phosphate.

25.37 No. Arginase is an enzyme found only in the livers of terrestrial animals.

25.39 Phenylketonuria, which causes defects in the central nervous system, is the result of a defect in the enzyme that oxidizes phenylalanine to form tyrosine. The hydrolysis of aspartame in the intestine will produce phenylalanine, which phenylketonurics must avoid.

25.41 False. Oxidative deamination of aspartate leads to the formation of pyruvate and ammonium ion. Ammonia cannot exist at physiological pH.

Chapter 26

Hormones and the Control of Metabolic Interrelations

26.1 The exchange of nutrients and waste products with the external environment is easily accomplished by simple diffusion in a single cell. However, there is a high-energy cost for maintaining a constant internal environment in the face of changing external environmental conditions.

26.3 Foods are enzymatically degraded to low-molecular-mass components to prepare them for absorption in the gut.

26.5 They bind to protein receptors on cell membranes and cause a second messenger to be activated in the cytosol, which initiates a variety of enzymatic cascades.

26.7 A zymogen is an inactive enzyme precursor.

26.9 They are transported across intestinal cells into the bloodstream and directly into the portal circulation.

26.11 Gastrin is secreted in the stomach in response to the entry of protein and stimulates the secretion of pepsinogen and HCl.

26.13 The biological value of a protein quantifies the nutritional value of a protein. If a given protein provides all the required amino acids in the proper proportions and all are released upon digestion and absorbed, then the protein is said to have a biological value of 100.

26.15 Fatty acids containing more than one unsaturated bond past carbon 9 of a saturated chain, such as linoleic and linolenic acids, cannot be synthesized by humans. Prostaglandins, a family of lipid-soluble organic acids that are regulators of hormones, are synthesized by mammals from arachidonic acid (a polyunsaturated fatty acid that mammals can synthesize with the use of dietary polyunsaturated fatty acids of plant origin as precursors).

26.17 A dietary deficiency disease is caused by a deficiency in a factor essential to cellular function, which can only be obtained through the diet.

26.19 The brain uses mostly glucose and some 3-hydroxybutyrate for energy needs.

26.21 A large electrical potential across its cell membranes and the ability to transmit an electrical signal from cell to cell are two unique physiological properties of the brain.

26.23 The distributions of sodium and potassium ions across nerve-cell membranes are:

$Na^+_{in} = 15$ mM, $Na^+_{out} = 150$ mM; $K^+_{in} = 125$ mM, $K^+_{out} = 10$ mM

26.25 The brain synthesizes a large number of hormones and hormone-releasing agents that affect distant organs and tissues.

26.27 There are virtually no mechanisms for storing energy in heart cells.

26.29 Because the heart prefers fatty acids as a source of energy, it must depend on the citric acid cycle for its ATP.

26.31 Glycogen is the principal energy source of active muscle.

26.33 Lipases on adipocyte cell surfaces hydrolyze the triacylglycerols of the chylomicrons, allowing the resulting fatty acids to enter the cells.

26.35 Glucagon stimulates lipases within adipocytes.

26.37 Glycerol that is formed from the hydrolysis of triacylglycerols eventually reaches the liver to contribute to gluconeogenesis.

26.39 Respiration of kidney cells is used to produce the ATP necessary to effect the assymetric distribution of Na^+ ions around the kidney tubules

26.41 Increasing the ammonia concentration of the urine, thereby increasing the amount of ammonium ion excreted, can eliminate excessive amounts of hydrogen ion.

26.43 The liver possesses glucokinase, an enzyme able to phosphorylate glucose at the high concentrations present after a meal.

26.45 In the liver, ammonia is converted into urea.

26.47 Acetoacetate and 3-hydroxybutyrate are the ketone bodies.

26.49 The liver synthesizes the bile acids from cholesterol that are used in the intestinal digestion of dietary lipids.

26.51 The absorptive state is the condition of the body immediately after a meal, when the gastrointestinal tract is full.

26.53 Insulin is present in high concentrations in the absorptive state.

26.55 Glucagon is the hormone that is present in the postabsorptive state.

26.57 In the postabsorptive state, blood glucose is supplied by glycogenolysis. When starvation begins, glycogen is unavailable and gluconeogenesis with the use of glucogenic amino acids from muscle provides the glucose.

26.59 It typically appears early in life and can be controlled by insulin replacement.

26.61 The liver maintains a supply of glycogen by gluconeogenesis with the use of glucogenic amino acids, which provides the glucose.

26.63 Because glucose is not available for energy production, fatty acid oxidation becomes the main source of ATP.

26.65 The cells present are erythrocytes, leukocytes, and platelets.

26.67 The presence of hemoglobin in eythrocytes increases the oxygen carrying capacity of blood.

26.69 2,3-Bisphosphoglycerate lowers the affinity of hemoglobin for oxygen.

26.71 Bicarbonate ion is formed in erythrocytes from CO_2, under the influence of the enzyme carbonic anhydrase.

26.73 A high concentration of CO_2 lowers the pH, and the Bohr effect enhances oxyhemoglobin dissociation.

26.75 In respiratory alkalosis, the blood pH rises to higher than normal levels because CO_2 is being eliminated faster than it is being formed by respiring cells.

26.77 A lipoprotein consists of a core of hydrophobic lipids surrounded by a shell of amphipathic lipids and proteins.

26.79 The proteins in lipoproteins solubilize lipids, direct specific lipoproteins to particular tissues, and activate enzymes that hydrolyze and unload lipids from lipoproteins.

26.81 Nitrogen balance is achieved when the intake of protein nitrogen is equal to the loss of nitrogen in the urine and feces.

26.83 This vitamin is not ordinarily essential, because it is synthesized in adequate amounts by intestinal bacterial flora. However, vitamin B_{12} is transported across the intestinal cell membrane as a complex with intrinsic factor; in persons who cannot synthesize this protein, the vitamin cannot enter the bloodstream through intestinal absorption and thus must be supplemented in the diet.

26.85 The liver uses excess glucose for the synthesis of fatty acids and cholesterol.

26.87 If it were synthesized as the active proteolytic enzyme chymotrypsin, it would digest the pancreas itself.

26.89 Dietary lipid in the form of chylomicrons is hydrolyzed at the surface of capillary membranes within muscle and adipose tissue. The free fatty acids then penetrate the cells of the tissue to be stored as triacylglycerol or oxidized for energy.

26.91 The protein contained in the wheat is difficult to extract because it is located within an indigestible husk.

APPENDIX

Solutions to *Organic and Biochemistry* Chapters 1 and 2

Chapter 1. The Properties of Atoms and Molecules

1.1 Proton: mass 1.00728 amu, charge = +1

Neutron: mass 1.00894 amu, charge = 0

Electron: mass 0.0005414 amu, charge = −1

1.2 Atomic number = 16; element is sulfur.

1.3 Atomic mass includes the mass of neutrons.

1.4

Protons	Neutrons	Electrons	Element
19	20	19	K
34	45	34	Se
6	6	6	C
11	12	11	Na

1.5 It would be a cation with a charge = +3.

1.6 It would be an anion with a charge = −1.

1.7

Element	Group	Period
Li	I	2
Na	I	3

K	I	4
Rb	I	5
Cs	I	6

1.8

Element	Group	Period
Si	IV	3
Ge	IV	4
As	V	4
Sb	V	5

1.9 Li, Na, K, Rb, and Cs are metals because they are all on the left side of the periodic table.

1.10 Si, Ge, As, and Sb are elements that have properties of both metals and nonmetals and are called metalloids.

1.11 Group I

1.12 Group VII

1.13 Group I. All elements possess one outer electron, ns^1.

1.14 Group VII. All elements possess seven outer electrons, ns^2np^5.

1.15 Group II. All elements possess two outer electrons, ns^2.

1.16 Group VIII. All elements possess eight outer electrons, ns^2np^6, except helium, $1s^2$.

1.17 True

1.18 By losing, gaining, or sharing electrons.

1.19 The formation of either ionic or covalent bonds.

1.20 Group I forms 1+ ions; Group VII forms 1– ions. For neutral compound, ratio = 1:1.

1.21 Group I forms 1+ ions; Group VI forms 2– ions. For neutral compound, ratio = 2:1.

1.22 Group II forms 2+ ions; Group VII forms 1– ions. For neutral compound, ratio = 1:2.

1.23 Group III forms 3+ ions; Group VI form 2– ions. For neutral compound, ratio = 2:3.

1.24 Calcium ion = Ca^{2+}. Any Group VI element forms a 2– ion. Therefore, to make a neutral compound, one calcium and one Group VI element must be included in the compound. The first three elements in Group VI are oxygen (O), sulfur (S), and selenium (Se). Thus, the compounds are CaO, CaS, and CaSe.

1.25 Aluminum ion = Al^{3+}. Any Group VI element forms a 2– ion. Therefore, to make a neutral compound, two aluminums and three Group VI elements must be included in the compound. The first three elements in Group VI are oxygen (O), sulfur (S), and selenium (Se). Thus, the compounds are Al_2O_3, Al_2S_3, and Al_2Se_3.

1.26 Aluminum ion = Al^{3+}. Any Group VII element forms a 1– ion. Therefore, to make a neutral compound, one aluminum and three Group VII elements must be included in the compound. The first four elements in group VII are Fluorine (F), chlorine (Cl), bromine (Br), iodine (I). Thus, the compounds are AlF_3, $AlCl_3$, $AlBr_3$, and AlI_3.

1.27 Procedure for drawing Lewis dot structures:

(i) Place atoms and connect bonds.

(ii) Determine how many valence electrons are available from all atoms.

(iii) Subtract two electrons for each bond drawn in part i.

(iv) Distribute remaining electrons among the atoms. Result should be that every atom has eight electrons associated with it from either free pairs of electrons or covalent bonds.

CCl_4

(i)

```
        Cl
        |
  Cl — C — Cl
        |
        Cl
```

(ii) 4 electrons from the carbon + (7 electrons from each chlorine × 4) = 32 e^-.

(iii) Covalent bonds use 8 electrons → 32 – 8 = 24 electrons remain.

(iv) Distribute 24 electrons among four Cl atoms (6 each) to give each Cl atom 8 electrons.

```
        :Cl:
         |
  :Cl — C — Cl:
         |
        :Cl:
```

Check that each atom has 8 electrons associated with it. ---- ✓

1.28 Use steps i through iv in the solution to Exercise 1.27:

(i)

$$\begin{array}{c} Cl \\ | \\ H-C-Cl \\ | \\ Cl \end{array}$$

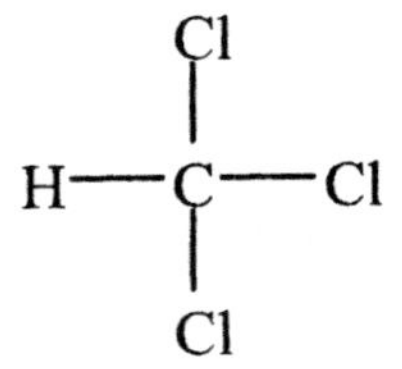

(ii) 4 electrons from the carbon + (7 electrons from each chlorine × 3) + 1 electron from the hydrogen = 26 e⁻.

(iii) Covalent bonds use 8 electrons → 26 – 8 = 18 electrons remain.

(iv) Distribute 18 electrons among three Cl atoms (6 each) to give each Cl atom 8 electrons.

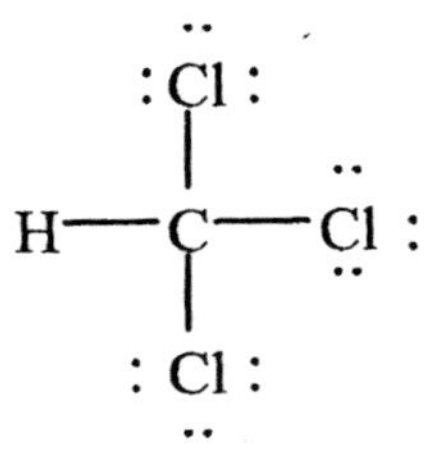

$$\begin{array}{c} :\ddot{Cl}: \\ | \\ H-C-\ddot{Cl}: \\ | \\ :\ddot{Cl}: \end{array}$$

Check that each atom has 8 electrons associated with it. ---- ✓

1.29 Use steps i through iv in the solution to Exercise 1.27:

(i)

$$H-S-H$$

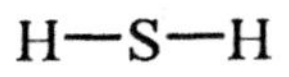

(ii) 6 electrons from the sulfur + (1 electron from each hydrogen × 2) = 8 e⁻.

(iii) Covalent bonds use 4 electrons → 8 – 4 = 4 electrons remain.

(iv) Distribute 4 electrons among H and S. First put 4 electrons (two free pairs) on the sulfur to give it an octet.

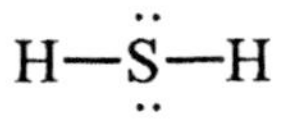

$$H-\ddot{\underset{..}{S}}-H$$

Check that each atom has 8 electrons associated with it. ---- ✓

1.30 Use steps i through iv in the solution to Exercise 1.27:

(i)

$$Cl-N-Cl$$
$$|$$
$$Cl$$

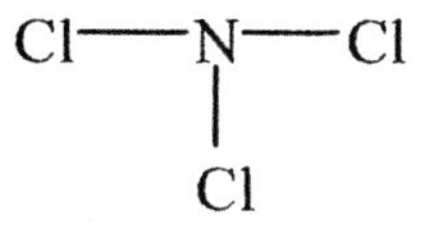

(ii) 5 electrons from the oxygen + (7 electrons from each chlorine × 3) = 26 e⁻.

(iii) Covalent bonds use 6 electrons → 26 − 6 = 20 electrons remain.

(iv) Distribute 20 electrons among three chlorines and one nitrogen. First put 2 electrons (one free pair) on the nitrogen to give it an octet. That leaves 18 electrons; distribute these 18 electrons among three Cl atoms (6 each) to give each Cl atom 8 electrons.

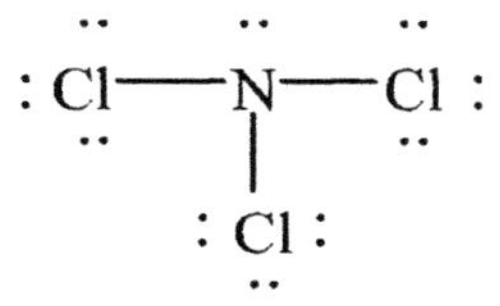

$$:\ddot{C}l-\ddot{N}-\ddot{C}l:$$
$$|$$
$$:\ddot{C}l:$$

Check that each atom has 8 electrons associated with it. ---- ✓

1.31 Use steps i through iv in the solution to Exercise 1.27:

(i)

$$O-C-O$$

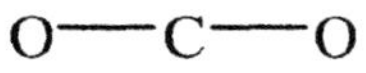

(ii) 4 electrons from the carbon + (6 electrons from each oxygen × 2) = 16 e⁻.

(iii) Covalent bonds use 4 electrons → 16 − 4 = 12 electrons remain.

(iv) Distribute 12 electrons among C and O. First put 4 electrons (two free pairs) on the carbon to give it an octet. That leaves 8 electrons; distribute these 8 electrons among two O atoms (4 each) to give:

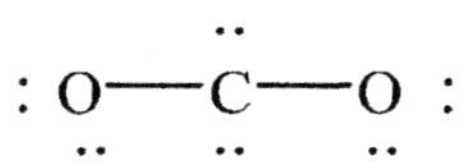

$$:\ddot{O}-\ddot{C}-\ddot{O}:$$

Check that each atom has 8 electrons associated with it. ---- **NO**

Each oxygen has only 6 electrons. To remedy this, we need to incorporate multiple (double) bonds. When there are not enough electrons to fill each atom to an octet, changing single bonds to double bonds will result in more sharing of electrons.

Try again:

(i) Make bonds between C and O double bonds:

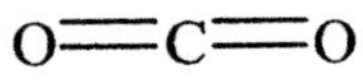

$$O=C=O$$

(ii and iii) Covalent bonds use 8 electrons → 16 – 8 = 8 electrons remain.

(iv) Distribute remaining electrons between the two oxygens because the carbon has 8 electrons associated with it.

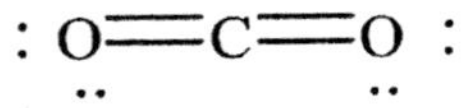

$$:\!\ddot{O}=C=\ddot{O}\!:$$

Check that each atom has 8 electrons associated with it. ---- ✓

1.32 Use steps i through iv in the solution to Exercise 1.27:

(i)

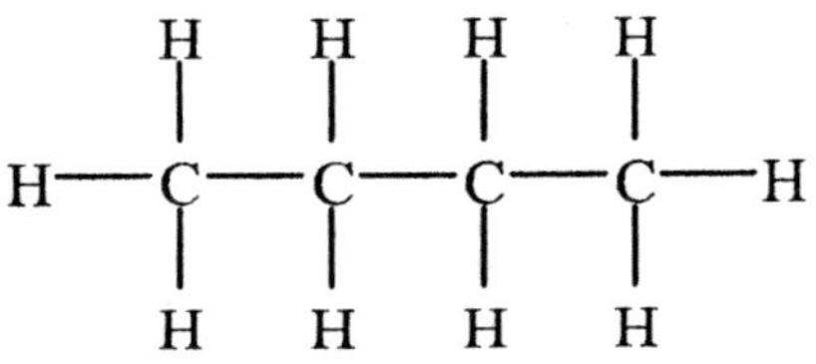

```
      H   H   H   H
      |   |   |   |
  H — C — C — C — C — H
      |   |   |   |
      H   H   H   H
```

(ii) (4 electrons from the carbon × 4) + (1 electron from each hydrogen × 10) = 26 e⁻.

(iii) Covalent bonds use 26 electrons → 26 – 26 = 0 electrons remain.

(iv) No more free electrons; this structure is the Lewis dot structure:

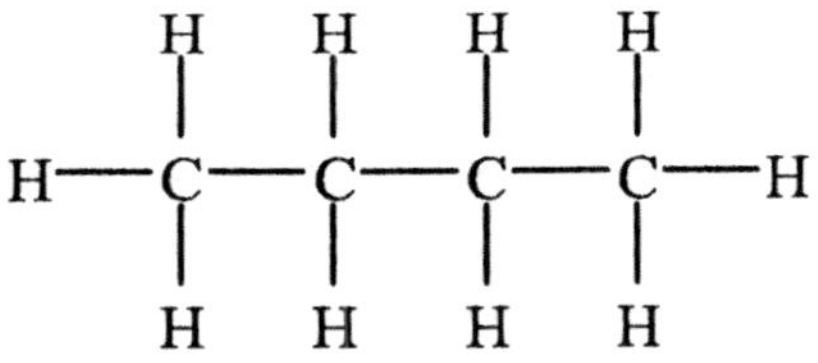

```
      H   H   H   H
      |   |   |   |
  H — C — C — C — C — H
      |   |   |   |
      H   H   H   H
```

Check that each atom has 8 electrons associated with it. ---- ✓

1.33 VSEPR provides information on the three-dimensional structure of a molecule by taking into account the fact that electrons have a negative charge. Therefore each pair of electrons in an atom "wants" to be as far away from all other pairs of electrons as poaaible. In this theory, "pairs of electrons" refers to either nonbonded pairs of electrons or covalent bonds. Furthermore, single, double, and triple bonds all count as a single "pair of electrons."

Therefore, before using VSEPR to determine the three-dimensional arrangement of the bonds and nonbonded electron pairs, we must first determine the Lewis dot structures. We will use the procedure outlined in the solution to Exercise 1.27.

(i)

(ii) 3 electrons from the boron atom + (7 electrons from each fluorine × 4) + 1 extra electron from the (–) charge = 32 e⁻.

(iii) Covalent bonds use 8 electrons → 32 – 8 = 24 electrons remain.

(iv) Distribute 24 electrons among four F atoms (6 each) to give:

$$:\ddot{\underset{..}{F}}:$$
$$:\ddot{\underset{..}{F}}\!-\!B\!-\!\ddot{\underset{..}{F}}:$$
$$:\ddot{\underset{..}{F}}:$$

Therefore, the central atom is boron and it has four bonds, which equals 4 "pairs of electrons," around it. These four bonds must arrange so that they are as far away from each other as possible. This arrangement is accomplished by having the four fluorine atoms situated at the corners of a tetrahedron, with the boron atom at the center.

1.34   From 1.29, the Lewis dot structure of H_2S is

$$H\!-\!\ddot{\underset{..}{S}}\!-\!H$$

Therefore, the central atom is sulfur and it has two bonds and two nonbonded pairs of electrons, which equals four "pairs of electrons," around it. These four "pairs" must arrange so that they are as far away from each other as possible. This arrangement is accomplished by having the two fluorine atoms and the two nonbonded pairs situated at the corners of a tetrahedron. Thus, the H–S–H bonds will form a bent molecule.

1.35   The Lewis dot structures are taken from the solutions to Exercises 1.27, 1.33, and 1.31 for parts a, b, and c, respectively.

(a)

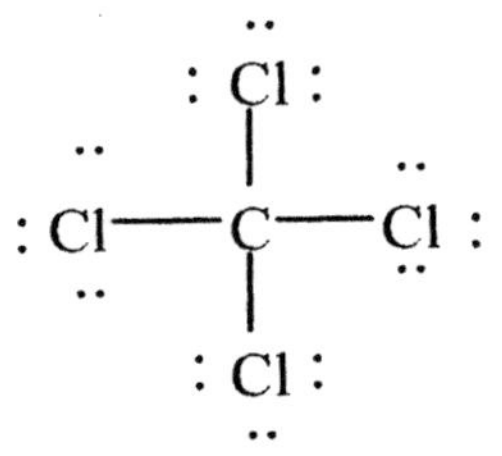

$$\text{:Cl—C(—Cl:)(—Cl:)—Cl:}$$

From this structure and VSEPR, CCl₄ forms a tetrahedron with each C–Cl bond going from the center to the corner. Even though each C–Cl bond is polar, this arrangement results in a cancellation of these polar contributions to give a molecule with no dipole moment.

(b)

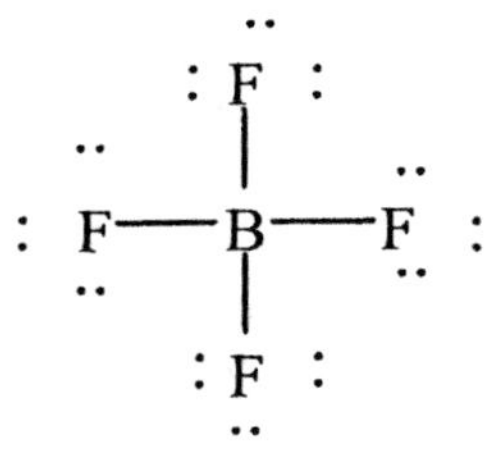

$$\text{:F—B(—F:)(—F:)—F:}$$

From this structure and VSEPR, BF₄ forms a tetrahedron with each B–F bond going from the center to the corner. Even though each B–F bond is polar, this arrangement results in a cancellation of these polar contributions to give a molecule with no dipole moment.

(c)

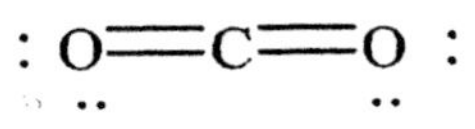

$$:O{=}C{=}O:$$

From this structure and VSEPR, CO₂ forms a linear molecule with each C=O 180° from the other. Even though each C=O bond is polar, this arrangement results in a cancellation of these polar contributions to give a molecule with no dipole moment.

1.36 The Lewis dot structures are taken from Exercise 1.28, 1.30, and Example 1.8 for parts a, b, and c, respectively.

(a)

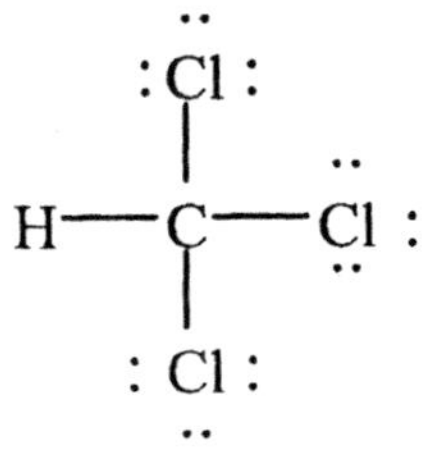

$$\text{H—C(—Cl:)(—Cl:)—Cl:}$$

From this structure and VSEPR, CHCl₃ forms a tetrahedron with each C–Cl (or C–H) bond going from the center to the corner. Because each C–Cl and C–H bond is polar and of different strengths, this arrangement will not cancel out the polar contribution of each bond and therefore results in a molecule with a dipole moment.

(b)

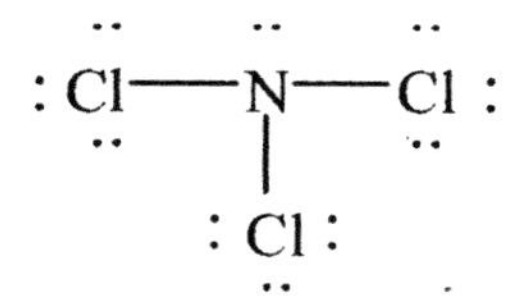

$$:\ddot{\text{C}}\text{l}\!-\!\overset{\cdot\cdot}{\underset{\underset{:\ddot{\text{C}}\text{l}:}{|}}{\text{N}}}\!-\!\ddot{\text{C}}\text{l}:$$

From this structure and VSEPR, NCl₃ forms a trigonal pyramid with each N–Cl bond going from the apex to the bottom. Because each N–Cl bond is polar, this arrangement results in a molecule with a dipole moment.

(c)

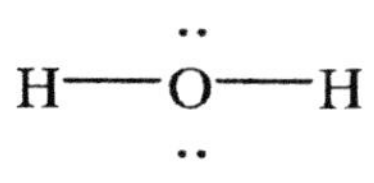

$$\text{H}\!-\!\overset{\cdot\cdot}{\underset{\cdot\cdot}{\text{O}}}\!-\!\text{H}$$

From this structure and VSEPR, H₂O forms a bent molecule. Because each H–O bond is polar, this arrangement results in a molecule with a dipole moment

1.37 The four C–Cl polar bonds are arranged symmetrically (tetrahedrally) with carbon at the center of the tetrahedron. The symmetric arrangement of dipoles results in a net dipole moment of zero.

1.38 Carbon dioxide is a linear molecule. The C–O dipoles are directly opposed to each other, so the net dipole moment is zero.

1.39

	London force	Dipole–dipole	Hydrogen bond
CH_4	Yes	No	No
$CHCl_3$	Yes	Yes	No
NH_3	Yes	Yes	Yes

1.40

	London force	Dipole–dipole	Hydrogen bond
CCl_4	Yes	No	No
CHF_3	Yes	Yes	No
H_2O	Yes	Yes	Yes

1.41 Acetic acid molecules form hydrogen bonds both among themselves and with water. The attractive forces between acetic acid and water are similar, and therefore acetic acid will dissolve in water.

1.42 Pentane molecules can interact only by London forces, whereas water can form hydrogen bonds. The attractive forces between pentane molecules and water are very different and will therefore not form a solution.

1.43 They have virtually identical chemical properties owing to similar outer-shell electronic structure.

1.44 Sulfur is a nonmetal.

1.45 Calcium is a metal.

1.46 They are inert and do not readily react to form compounds, because each has a full outer electron shell.

1.47 In chemical reactions, atoms tend to attain the noble-gas outer-shell configuration of eight electrons.

1.48 The statement is incorrect; no more than eight elements have equal numbers of protons, electrons, and neutrons: He, C, N, O, Ne, Si, S, and Ca.

1.49 The chemical properties are identical, but the nucleus of carbon-12 contains six neutrons and that of carbon-13 contains seven neutrons.

1.50 Total percent = 100; isotope at 26 amu = X%; isotope at 25 amu = (100 – X)%

$$100\left[\frac{X}{100}(26) + \frac{100-X}{100}(25) = 25.6\right]$$

26X – 25X + 2500 = 2650

X = 60, 100 – X = 40

Isotope at 26 amu = 60%   isotope at 25 amu = 40%

1.51   Use steps i through iv in the solution to Exercise 1.27:

(i)

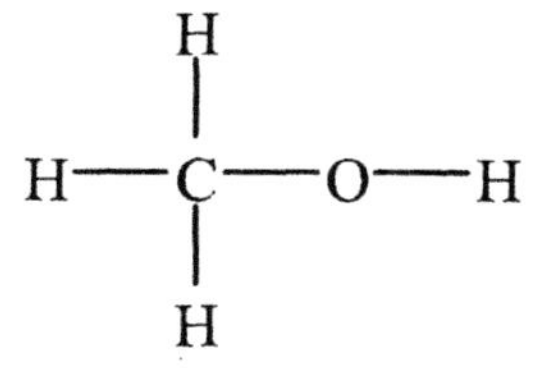

$$\begin{array}{c} H \\ | \\ H-C-O-H \\ | \\ H \end{array}$$

(ii) 4 electrons from the carbon + (1 electron from each hydrogen × 4) + 6 electrons from the oxygen = 14 e$^-$.

(iii) Covalent bonds use 10 electrons → 14 − 10 = 4 electrons remain.

(iv) Inspection shows that the outer shell of carbon and of each hydrogen is full, whereas oxygen is still short. Therefore distribute 4 electrons to the O atom:

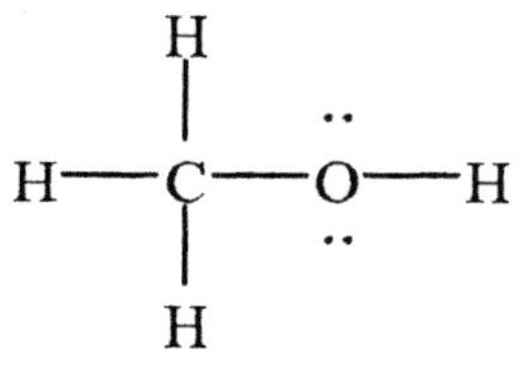

$$\begin{array}{c} H \\ | \\ H-C-\ddot{O}-H \\ | \\ H \end{array}$$

Check that each atoms has 8 electrons associated with it. ---- ✓

1.52   London dispersion forces are the result of temporary dipoles. A larger number of electrons in a molecule results in a larger temporary dipole. Therefore, the greater the molecular mass, the stronger the London forces between molecules.

Chapter 2. Chemical Change

2.1 From Table 2.1:

(a) AgCl is not soluble in H$_2$O; therefore a precipitate forms:

$$Ag^+ \text{ (aq)} + Cl^- \text{ (aq)} \rightarrow AgCl \text{ (s)}$$

(b) Mg(OH)$_2$ is not soluble in H$_2$O; therefore a precipitate forms:

$$Mg^{2+} \text{ (aq)} + 2\ OH^- \text{ (aq)} \rightarrow Mg(OH)_2 \text{ (s)}$$

2.2 (a) PbCl$_2$ is not soluble in H$_2$O; therefore a precipitate forms:

$$Pb^{2+} \text{ (aq)} + 2Cl^- \text{ (aq)} \rightarrow PbCl_2 \text{ (s)}$$

(b) Ca$_3$(PO$_4$)$_2$ is not soluble in H$_2$O; therefore a precipitate forms:

$$3\ Ca^{2+} \text{ (aq)} + PO_4^{3-} \text{ (aq)} \rightarrow Ca_3(PO_4)_2 \text{ (s)}$$

2.3 (a) BaSO$_4$ is not soluble in H$_2$O; therefore a precipitate forms:

$$Ba^{2+} \text{ (aq)} + SO_4^{2-} \text{ (aq)} \rightarrow BaSO_4 \text{ (s)}$$

(b) Al(OH)$_3$ is not soluble in H$_2$O; therefore a precipitate forms:

$$Al^{3+} \text{(aq)} + OH^- \text{(aq)} \rightarrow Al(OH)_3 \text{ (s)}$$

2.4 (a) AgBr is not soluble in H$_2$O; therefore a precipitate forms:

$$Ag^+ \text{(aq)} + Br^- \text{(aq)} \rightarrow AgBr \text{ (s)}$$

(b) There is no reaction, because all possible salts are soluble.

2.5 For reaction a to take place, the two reactants must have kinetic energies (K.E.) greater than 24 kJ; for reaction b to take place the two reactants must have kinetic energies greater than 53 kJ. At a given temperature, the percentage of molecules that have K.E. > 24 kJ is greater than the percentage that have K.E. > 53 kJ. Therefore, the rate of reaction a is greater than the rate of reaction b.

2.6 The enzyme catalase is a catalyst that lowers the activation energy of the reaction, which in turn increases the reaction rate at a given temperature.

2.7 $\Delta H_{reaction} = E_{forward} - E_{back}$

$-21 \text{ kJ} = 37 \text{ kJ} - E_{back}$

$E_{back} = 58 \text{ kJ}$

2.8 $\Delta H_{reaction} = E_{forward} - E_{back}$

$12 \text{ kJ} = -46 \text{ kJ} - E_{back}$

$E_{back} = -58 \text{ kJ}$

2.9 No. The reaction takes place in an open system, so the CO_2 escapes to the atmosphere. Therefore, the CO_2 that is formed cannot react with CaO to allow the reverse reaction to take place. Therefore, the system can never come to equilibrium.

2.10 No. The reaction takes place in an open system, so the CO_2 escapes to the atmosphere. Therefore, the CO_2 that is formed cannot react with H_2O and $CaCl_2$ to allow the reverse reaction to take place. Therefore, the system can never come to equilibrium.

2.11 Forward reaction: $CaCO_3$ (s) → CaO (s) + CO_2 (g)

Reverse reaction: CaO (s) + CO_2 (g) → $CaCO_3$ (s)

2.12 Forward reaction: CaO (s) + CO_2 (g) → $CaCO_3$ (s)

Reverse reaction: $CaCO_3$ (s) → CaO (s) + CO_2 (g)

2.13 $K_{eq} = \dfrac{[HI]^2}{[H_2][I_2]} = \dfrac{(0.27)^2}{(0.86)(0.86)} = 0.099$

2.14 $K_{eq} = [NH_3][HCl] = [3.7 \times 10^{-3}][3.7 \times 10^{-3}] = 1.38 \times 10^{-5}$

2.15 Le Chatelier's principle states that, when a reactant or product is removed from a reaction at equilibrium, the equilibrium shifts to replace the compound that was removed.

$$N_2 \text{ (g)} + 3 H_2 \text{ (g)} \leftrightarrow 2 NH_3$$

If NH_3 is removed, the reaction shifts to replace NH_3, which means that it shifts to the products.

2.16 Le Chatelier's principle states that, when the temperature is raised in a reaction at equilibrium, the equilibrium shifts to lower the temperature. In this reaction, that means that the reaction shifts to the reactants.

2.17 Le Chatelier's principle states that, when a reactant or product is added to a reaction at equilibrium, the equilibrium shifts to use up the compound that was added. In this reaction, when Cl^- is added, the reaction shifts to use up more Cl^-; the reaction will shift to the products, which will result in a color shift to blue.

2.18 Le Chatelier's principle states that when a reactant or product is added to a reaction at equilibrium, the equilibrium shifts to use up the compound that was added. In this reaction, when Fe^{3+} is added, the reaction shifts to use up more Fe^{3+}; the reaction will shift to the products which will result in a color shift to red.

2.19 $COCl_2\ (g) \leftrightarrow CO\ (g) + Cl_2\ (g)$

2.20 $CH_4\ (g) + H_2O\ (g) \leftrightarrow CO\ (g) + 3\ H_2\ (g)$

2.21 $PCl_3\ (g) + Cl_2\ (g) \leftrightarrow PCl_5\ (g)$

2.22 $3\ O_2\ (g) \leftrightarrow 2\ O_3\ (g)$

2.23 The concentrations of pure liquids and solids do not appear in K_{eq}; therefore:

$$K_{eq} = \frac{[NO_2][NO_3]}{[N_2O_5]}$$

2.24 The concentrations of pure liquids and solids do not appear in K_{eq}; therefore:

$$K_{eq} = \frac{[N_2][H_2O]^2}{[NO]^2[H_2]^2}$$

2.25 The ion product, $K_w = [H_3O^+] \times [OH^-] = (1.00 \times 10^{-7}) \times (1.00 \times 10^{-7}) = 1.00 \times 10^{-14}$

2.26 In pure water $[H_3O^+] = [OH^-]$ and $K_w = [H_3O^+] \times [OH^-]$. Therefore, $K_w = [OH^-]^2 = [H_3O^+]^2 = 2.51 \times 10^{-14}$. $[H_3O^+] = [OH^-] = (K_w)^{1/2} = 1.58 \times 10^{-7}$.

2.27 A strong base is ionized 100% in aqueous solution. Two examples are NaOH and LiOH.

2.28 A strong acid is ionized 100% in aqueous solution. Two examples are HCl and H_2SO_4.

2.29 0.30 M HCl solution. HCl is a strong acid, which means that every HCl molecule dissociates into an H_3O^+ ion and a Cl^- ion. Thus, $[H_3O^+] = [Cl^-] = 0.30$ M.

2.30 0.28 M $HClO_3$ solution. $HClO_3$ is a strong acid, which means that every $HClO_3$ molecule dissociates into an H_3O^+ ion and a ClO^- ion. Thus, $[H_3O^+] = [ClO^-] = 0.28$ M.

2.31 The pH of pure water at 25°C is 7.00.

2.32 The pH of water at 37°C is 6.80. The equilibrium reaction between two water molecules to form H_3O^+ and OH^- changes with temperature, and, more H_3O^+ is formed at 37°C than at 25°C, where the pH is 7.00.

2.33 $pH = -\log [H_3O^+] = -\log (0.01) = 2.0$

2.34 $pH = -\log [H_3O^+] = -\log (0.00263) = 2.58$

2.35 A weak acid is one that incompletely dissociates in water. Examples are given in Table 2.4: acetic acid and carbonic acid.

2.36 A weak base is one that incompletely dissociates in water. Examples are given in Table 2.4: aniline and morphine.

2.37 (a) An acid is a compound that donates a proton; its conjugate base is the compound that is formed upon loss of a hydrogen. Both HNO_2 and H_3O^+ give up a proton to form NO_2^- and H_2O, respectively. Therefore HNO_2 and NO_2^- are one conjugate acid-base pair, and H_3O^+ and H_2O are another.

(b) An acid is a compound that donates a proton; its conjugate base is the compound that is formed upon loss of a hydrogen. Both $H_2PO_4^-$ and H_3O^+ give up a proton to form HPO_4^{2-} and H_2O, respectively. Therefore $H_2PO_4^-$ and HPO_4^{2-} are one conjugate acid-base pair, and H_3O^+ and H_2O are another.

2.38 (a) An acid is a compound that donates a proton; its conjugate base is the compound that is formed upon loss of a hydrogen. Both HCOOH and NH_4^+ give up a proton to form $HCOO^-$ and NH_3, respectively. Therefore HCOOH and $HCOO^-$ are one conjugate acid-base pair, and NH_4^+ and NH_3 are another.

(b) An acid is a compound that donates a proton; its conjugate base is the compound that is formed upon loss of a hydrogen. Both H_2O and H_3O^+ give up a proton to form OH^- and H_2O, respectively. Therefore H_2O and OH^- are one conjugate acid-base pair, and H_3O^+ and H_2O are another.

2.39 Because the presence of the conjugate acid allows the solution to neutralize the addition of a base while the presence of the conjugate base allows the solution to neutralize the addition of a base. Because it is a conjugate acid-base pair, the solution remains an a constant pH.

2.40 $pH = pK_a + \log\left(\dfrac{proton\ acceptor}{proton\ donor}\right) = \log\left(\dfrac{0.0080}{0.0060}\right) = 4.76 + 0.125 = 4.89$

2.41 $pH = pK_a + \log\left(\dfrac{proton\ acceptor}{proton\ donor}\right) = \log\left(\dfrac{0.070}{0.070}\right) = 3.75 + 0 = 3.75$

2.42 $pH = pK_a + \log\left(\dfrac{proton\ acceptor}{proton\ donor}\right) = \log\left(\dfrac{0.075}{0.050}\right) = 7.20 + 0.18 = 7.38$

2.43 $pH = pK_a + \log\left(\dfrac{proton\ acceptor}{proton\ donor}\right) = \log\left(\dfrac{0.050}{0.075}\right) = 7.20 - 0.176 = 7.02$

2.44 (a) 0.0031 M HNO_3 solution. HNO_3 is a strong acid, which means that every HNO_3 molecule dissociates into an H_3O^+ ion and an NO_3^- ion. Thus, $[H_3O^+] = 0.0031$ M.

$$\text{pH} = -\log [H_3O^+] = -\log (0.0031) = 2.50$$

(b) 1.0 M HCl solution. HCl is a strong acid, which means that every HCl molecule dissociates into an H_3O^+ ion and a Cl^- ion. Thus, $[H_3O^+] = 1.0$ M.

$$\text{pH} = -\log [H_3O^+] = -\log (1.0) = 0.0$$

2.45 (a) 0.0069 M HI solution. HI is a strong acid, which means that every HI molecule dissociates into an H_3O^+ ion and an I^- ion. Thus, $[H_3O^+] = 0.0069$ M.

$$\text{pH} = -\log [H_3O^+] = -\log (0.0069) = 2.2$$

(b) 0.019 M HBr solution. HBr is a strong acid, which means that every HBr molecule dissociates into an H_3O^+ ion and a Br^- ion. Thus, $[H_3O^+] = 0.019$ M.

$$\text{pH} = -\log [H_3O^+] = -\log (0.019) = 1.7$$

2.46 The combination of a weak acid and the salt of its conjugate base is a buffer solution. The pH of a buffer is:

$$\text{pH} = pK_a + \log\left(\frac{\text{proton acceptor}}{\text{proton donor}}\right) = \log\left(\frac{0.075}{0.050}\right) = 6.35 + 0.125 = 6.53$$

2.47 The combination of a weak acid and the salt of its conjugate base is a buffer solution. The pH of a buffer is:

$$\text{pH} = pK_a + \log\left(\frac{\text{proton acceptor}}{\text{proton donor}}\right) = 8.2 = 7.2 + x$$

$$x = 1.0 = \log\left(\frac{\text{proton acceptor}}{\text{proton donor}}\right) \rightarrow HPO_4^{2-}/H_2PO_4^- = 10/1$$

2.48 $\text{pH} = -\log [H_3O^+] = 3.20 \rightarrow [H_3O^+] = 6.3 \times 10^{-4}$

HCl is a strong acid, which means that every HCl molecule dissociates into an H_3O^+ ion and a Cl^- ion. Thus, $[H_3O^+] = [HCl] = 6.3 \times 10^{-4}$ M.

2.49 (a) HCl is a strong acid, which means that every HCl molecule dissociates into an H_3O^+ ion and a Cl^- ion. Thus, $[H_3O^+] = [HCl] = 0.00336$ M. It is also known that

$$[H_3O^+] \times [OH^-] = 1 \times 10^{-14} \rightarrow [OH^-] = \frac{1 \times 10^{-14}}{0.00336} = 2.97 \times 10^{-12}$$

$$\text{pH} = -\log[H_3O^+] = 2.47$$

(b) HNO_3 is a strong acid, which means that every HNO_3 molecule dissociates into an H_3O^+ ion and an NO_3^- ion. Thus, $[H_3O^+] = [HNO_3] = 0.0253$ M. It is also known that

$$[H_3O^+] \times [OH^-] = 1 \times 10^{-14} \rightarrow [OH^-] = \frac{1 \times 10^{-14}}{0.0253} = 3.98 \times 10^{-13}$$

$$pH = -\log[H_3O^+] = 1.60$$

2.50   A combination of a weak acid and the salt of its conjugate base is a buffer solution. Thus examples include acetic acid and potassium acetate, formic acid and sodium formate, and carbonic acid and sodium bicarbonate.

2.51   A catalyst is a chemical that increases the rate of a reaction but is not used up in the overall reaction. Therefore, NO is the catalyst.

2.52 $Mg(OH)_2$ (s) $\leftrightarrow$ Mg^{2+} (aq) + 2 OH^- (aq)

If H^+ is added to this reaction, it will react with OH^-:

$$H^+ + OH^- \leftrightarrow H_2O$$

The net result is that OH^- is lost from the first equilibrium. Thus, that equilibrium will respond by trying to replace that OH^- by pushing that reaction to the right.

If a system in an equilibrium state is disturbed, the system will adjust to neutralize that disturbance and restore the system to equilibrium.

The ionization of water is an equilibrium reaction:

$$H_2O \leftrightarrow H_3O^+ + OH^-$$

The fact that the ion product of water increases with temperature means that this equilibrium is shifted to the right (products) with an increase in temperature. According to Le Chatelier's principle, when an equilibrium is perturbed, it will shift to bring the equilibrium back to its initial conditions. Therefore, when the temperature is increased, this reaction will shift to the right, which lowers the temperature. A reaction that lowers the temperature as it goes toward it products is endothermic.

2.55   Ethanol is in excess; therefore it is considered to be the solvent, and water is the solute.

2.56 $V_1 M_1 = V_2 M_2$

$V_1 = ?$ $M_1 = 0.30$ M

$V_2 = 135$ mL $M_2 = 0.180$ M

$$V_1 = \frac{135 \text{ mL} \times 0.180 \text{ M}}{0.30 \text{ M}} = 81.0 \text{ mL of original solution needed.}$$

Therefore, she needs 81.0 mL of original solution to make 135 mL of the final solution. Because she has only 70 mL, she does not have enough.

2.57 $13.9 \text{ g} \times \dfrac{1 \text{ mol}}{95.3 \text{ g}} = 0.146$ mol; 0.146 mol/0.225 L = 0.650 M

2.58 The square brackets around chemical components in an equilibrium constant expression denote their molar concentrations at equilibrium.

2.59 A conjugate acid-base pair is a weak acid and the basic anion that results from its dissociation.